J. N. (John Newton) Stearns

Merry's Vook of Animals

J. N. (John Newton) Stearns

Merry's Vook of Animals

ISBN/EAN: 9783743324244

Manufactured in Europe, USA, Canada, Australia, Japa

Cover: Foto ©berggeist007 / pixelio.de

Manufactured and distributed by brebook publishing software
(www.brebook.com)

J. N. (John Newton) Stearns

Merry's Vook of Animals

THE HAPPY FAMILY.

MERRY'S BOOK

OF

ANIMALS.

EDITED BY

UNCLE MERRY.

NEW-YORK:

H. DAYTON, No. 36 HOWARD STREET.

INDIANAPOLIS, IND.: ASHER & CO.

1860.

J. J. REED, PRINTER & STEREOTYPER,
43 & 45 Centre Street.

CONTENTS.

ENGRAVINGS.

PREFACE.

W E have had many a pleasant "chat" with our young friends about matters and things in general, and some in particular. We now invite them, one and all, to a special chat about the animal creation—not a brutal chat, but a chat about brutes. Come, let us ramble in the fields, and in the woods, let us walk about the farm-yards, and peep into the menageries and museums, and see what we can find to interest us.

The animals—the beasts of the field and the forest—were created before man ; but they were brought to Adam to be named. He was their acknowledged lord. Adam must have known a great deal of the characters and habits of the animals, to be able to give them appropriate names. How did he become acquainted with them so soon ?

Noah, too, must have understood the animals very well, to be able to provide for so many, and take care of them a whole year, in his great floating menagerie.

Solomon is said, also, to have studied and written much about the beasts. The books he wrote about them were not preserved. They are all lost. How much would not the world now give to see one of those books. Great and wise men have always loved to study the works of God—the trees—the animals—the stars. There is no study more interesting than that of animals. We never can become acquainted with the whole of them, perhaps ; but we can always be learning something about them. New wonders will be always meeting our eyes, as we read and study. And we shall be constantly gaining new and enlarged ideas of the wisdom, power, and goodness of God.

The study of the animal creation is not only interesting, but very useful, to young persons. They ought to learn, as early as possible, all they can about the characters and habits, not only of those domestic animals which they have about them at home, but of those around in the fields and forests, with which they often have much to do. Without this knowledge they cannot derive as much advantage as they otherwise might do, from the tame and useful animals, nor guard themselves wisely from the injuries which the wild, or venomous might inflict upon them. It is by knowing what animals are, and observing their habits, that men are able to manage them, train them to be submissive and useful, and turn them to profitable account. It is hoped that all our young friends will become interested in this study, and improve every opportunity to pursue it. In the present little work, we may tell them some things which many of them know already. But we think there are few of them who will not learn something new, and some things that will amuse and please them. We think, too, that most of them will eagerly ask for " more."

THE SHEPHERD BOY.

THERE was once a little boy named Dick. He took care of his father's sheep. His father was a farmer in the neighborhood of Paris. One day, a little boy, one of his schoolmates, came to pass the day with him on the mountain.

This was little Albert. He was very fond of Dick, and he was glad that his parents had given him permission to pass a day with his friend, and see how he took care of his sheep.

After they had walked about and played till they were tired, they sat down under a shady tree to rest, and Dick told Albert a story of what happened to him when he first began to look after the sheep.

One day, when his father thought he had been a particularly good boy, he gave him a little lamb for his own, and said to him, " Dick, you have taken such good care of my sheep, that it is but fair I should repay you. This lamb is your own property—put a collar and bell about it, and take care not to let it get lost."

Dick thanked his father. He was delighted with his little lamb—he caressed it—and it shared his luncheon with him.

One day, when Dick as usual drove his flock to pasture on the mountain, after having eaten a part of the provision which he had in his basket, he fell asleep. The poor child did not look forward to the misfortunes which threatened him. While he was asleep, Ba-ba, for that was the name he had given his favorite, in browsing and running about, got over to the other side of the mountain. It was lost—its cries could not be heard—it looked for its master, and the more it sought him, the farther it strayed from the right way. Poor Dick awoke, he called Ba-ba, but it did not come ; he rubbed his eyes to try to discover it, but all his searching was in vain ; his misfortune was certain. " Oh dear," cried he, " what shall I tell my father, who has given me this lamb ? how careless he will think I have been of his present, which I really prized so highly. And you, poor Ba-ba, what can have become of you ! You will die of hunger, and it will be my fault." And poor Dick could not help crying bitterly.

Suddenly he saw an old woman approaching him, who could scarcely walk, she was so tired. Dick was so full of his own trouble, that he did not at first pay her much attention ; but presently she spoke to him, and said, " Good-day, my good shepherd, you seem afflicted, and I am in trouble myself, for I have lost my way ; I have been traveling for six hours, and I have had no breakfast, and it is so hot that I am dying with thirst."

On hearing these words Dick went to his basket. " There, good mother," said he, " take this piece of cake and these cherries ; I am very glad I had not eaten up all my luncheon—wait a moment."

He then ran to a brook which was at the foot of the hill, and came back out of breath, with his little tin mug full of water, and he came very gently that he need not spill it.

" How kind and good you are !" said she, when
she had drank it.   Now tell me what you were crying for when I came up to you ?"

Then Dick told her his trouble.

" My son," said the old woman, "dry your tears,
for I know where your lamb is, and ——"

" Is it possible," cried Dick, hugging the old woman, " is it possible that I shall see again my poor
little Ba-ba ?   Tell me quickly where she is, that I
may run after her.   Ah, my good mother, how
much I am obliged to you ; where is it, where is my
poor little lamb ?"

" My son," said the old woman, " I will lead you
there, but wait till I am a little rested."

" But tell me, mother, only tell me, and I can go
myself."

" No, no, I will go with you."

Dick was much vexed at being obliged to wait,
but he knew how to show the proper respect to old
age.   He did not insist any further, but sat down
near her and waited till she was ready to set out.

" Here, my son," said she, " I should be glad to
make acquaintance with you ; your kind heart in-
terests me.   I am very old—I know a great many
things, and I may be useful to you in the course of
your life.   What is your father's name ?"

" Michael."

" Ah, I know him ; he is a good man ; and your
name ?"

"Dick, ma'am. But, my good ma'am, it seems to me that you do not look hot now, and we might set out."

LOST ON THE MOUNTAIN.

"Have patience, little boy, you have better legs than mine."

"Pardon me, my good mother, I will wait as long as you please."

"Oh, well," said she, getting up, "I will make an exertion for your sake, you are so gentle and patient; give me your arm."

Then **she** led him to the place where she found the little lamb, which she had asked a peasant to keep for its owner. Dick took his dear lamb, in his arms, thanked the old woman over and over again, and invited her to go with him to his father's house.

The good old woman was so much pleased with Dick, that she took upon herself to teach him to read and write—and he was better educated than any of the neighboring boys, for they were employed most of the time in keeping their father's sheep. He did all he could to instruct his companions, so that they were much improved.

This story Dick told Albert, but not exactly in these words, but this was the true state of the case. Albert was greatly pleased with it, and Dick showed him his lamb, which had grown up into a stout sheep, and had furnished its master with more than one pair of stockings.

Dick continued to study all his leisure hours and improve himself, till at last he became the schoolmaster of his native village. He was a very great favorite of the papas **and** mammas, and all the old ladies; and was as much liked **by the** children as any schoolmaster could **be.**

THE PET GOAT.

ONE bleak day in February, as little Fannie was
returning home from school through the woods

she found a little kid, apparently but a day old. The ground was covered with snow, and the poor little creature was so benumbed by the cold that it could scarcely move. Fannie took it in her arms, and hastened home. Her mother gave her permission to keep it as a pet, and Fannie immediately got a basket of nice dry hay, and laid the kid in it. She then got it some milk, which it lapped up, and appeared to like very much, as it got up and frisked around the room. Fannie wished to give her pet a name, and she at last bestowed upon it the name of "Billy."

He became very fond of her, and her presence created a sort of sunshine to him; when she was with him, he would frisk about and appear very happy. He grew finely, and was soon large enough to do a great deal of mischief. He would nip off the buds of young roses in Fannie's garden, and soil her clothes by rubbing against her and trying to jump into her lap. In this way he would sometimes make his mistress angry, but she soon made friends with him again.

He was of a very pretty color, white, with black spots all about the head and neck, black feet, and a black nose. One day Fannie was very much frightened about her little pet, for as she was walking along in the meadow, with little Billy frisking along by her side, a large mastiff belonging to one of the neighbors came up and began to bark and worry him. Fannie screamed, and soon one of her brothers ran to the assistance of Billy. But before he could drive away the dog it had bitten Billy's leg and broken it. Fannie cried a great deal about it, but, with her brother's assistance, she got the poor creature home, and washed and dressed its leg. In the course of three weeks it was perfectly well.

Billy was very fond of following his mistress, and very often wished to go to school with her, but he was always driven home. One day, however, Billy thought he would go to school, whether his mistress was willing or no ; so he followed her at a distance, and she had been in school but a few minutes when she heard all the children laugh, and looking to find the cause of their mirth, what should she see but little Billy sitting by her side ! Of course he was sent home again.

A short time after this, Fannie's father sold his farm, and purchased another in a different state, and he said it would be too much trouble to carry Billy along with them ; but Fannie would not hear of any such thing as leaving her pet behind ; so after a great deal of persuasion, and many tears on her part, her father consented to take him along. The day they were to depart, Billy was nowhere to be found. The servants and Fannie's brothers searched every part of the house and yard, and, at last, found him snugly taking a nap on some straw in the coal-house. He was so dirty that Fannie hardly knew him. However, the only thing she could do was to wash him, and caution him not to go into such a place again.

Billy's horns had come out, and it would make you laugh to see him butt. Billy liked his new home very well. There was a nice garden in front of the house, and a large meadow behind it, through which ran a pleasant little brook, its borders covered with wild flowers ; and every pleasant morning you might see Fannie and Billy skipping along, Fannie picking wild flowers, and Billy, every now and then, nipping off a daisy. Here Billy had no ugly mastiff, to fear.

But he could not always be young. He has now grown old and sedate, and does not frisk and play as he used to. He still likes Fannie as well as ever, but he does not follow her as much. He likes to sit down in the sun, with his head down as if meditating on some great subject. He has a nice house and lives very comfortably. The last time I saw him he was quite gray ; his beard was long and grizzly, and his aspect quite venerable.

SIMPLIFYING A SPELLING LESSON.—"Spell cat," said a little girl of five, the other day, to a small one of only three. "I can't," was the reply. "Well, then, if you can't spell cat, spell kitten."

HARRY HATCHET'S DOG.

ESSIE," said Harry one morning at breakfast, " I'm to have a fine dog one of these days."

" You are to have," exclaimed Uncle Hiram ; " why I should think you already have all you can possibly need ; let me see—Watch is one."

" He is not mine, he's the house-dog ; you want him to keep watch according to his name," said Harry.

" Well, there's Topsey, and Rover—"

" Oh, they are only fit for pets for the girls," said Harry, " and then 'Sport'—he is a first-rate fellow, but he is not fit for every thing."

" Very well—what is your new dog to be then ?" asked Uncle.

" A first-rate, a No. 1 hunting dog."

" So," whistled Uncle Hiram, " your trip last year and sporting experience therein, has given you a taste for that amusement ?"

" Yes, indeed," exclaimed Harry ; " I was never so happy in my life ; and when I am more used to it, I shall do so much better, that I shall enjoy it

more—I was so free and independent—and, moreover, I'm a very good shot, I would have you
know."

THE PETS.

"But you have not told us where this wonderful
dog is to come from, Harry," said Jessie.

"Oh, I forgot," said Harry; "Clem Harding,
who was with us last year, promised him to me, I
took such a fancy to him—he was so very intelligent."

"Which? Clem, or the dog?" asked Uncle.

"The dog?" said Harry; "Clem is intelligent enough, everybody knows, without my telling it."

"That's what always puzzles me, father," said Edith, looking up earnestly; "dogs are intelligent, and seem almost to think; then why haven't they minds as well as men and women?"

"A great many wiser heads than yours have been puzzled with such questions," answered her father; "we cannot certainly say in what consists man's superiority—nor how far the reasoning faculty of animals reaches."

"They sometimes seem to think just as much as any one of us, and just as well, too," said Edith.

"That they do," cried Harry; "and *I* never saw any *dog* that could think as well and sensibly as Clem's 'Ranger;' that's what took me completely."

"Won't you tell us about him, then?" said Jessie.

"Oh, yes—nothing would please me better, for I want you all to have a liking for my new pet when he comes. Let me see, he did so many sensible things, that I don't know where to begin."

"Oh, anything—the first thing you think of," cried Jessie, eager to hear anything in the shape of a story.

"Well, once when we were trying to shoot some ducks, we left our hats, etc., quite a distance from the river, and crept through the reeds so that they

might not see us and fly away, before we had a
chance at them ; when we got there we fired, and
didn't do much execution, so we concluded to lie
still, and have another shot when the birds alighted
again. Of course we wanted our hats ; so Clem
sent Ranger back for them ; first he took Clem's

OUR ROVER.

and started to bring that, but Clem motioned him
back, and he understood that he must bring both.
You ought to have seen him work. First he would
take one, and then try to get the other up, but the
broad stiff brims wouldn't let him, and away would
go the first one ; once he almost succeeded. The
two were fairly in his mouth, but they would not
stay there. He stopped, and was evidently deep
in thought. He stood perfectly still for about a

minute, looking at the hats, and then took one hat,
put it inside of the other, pushed it down with his
paw, and in the most satisfied, triumphant manner
seized them and brought them to us. Don't you
call that thinking ?"

"Bravo, Ranger," cried Uncle Hiram, "he certainly can reason pretty well."

"When is this wonderful Ranger coming ?" asked
Lucy.

"In a day or two," said Harry, springing up and
catching Jessie in his arms, to give her a lesson in
waltzing, as he called it, swinging her rapidly
round.

"He'll beat all the dumb creatures about here,
for sense *I* mean. We shan't hear anything more
of the squirrel after that ;" and so saying he waltzed
himself out of the room.

"Oh, father, that poor little boy and his squirrel—don't you remember ?" said Lucy and Edith in
one breath.

"We ought to go and see them, certainly," said
Uncle Hiram. "You must make some appoint-
ment for me, and I will keep it."

"Shall it be this afternoon, then, father ?" said
Edith, who could never wait long for any pleasure.

"Yes, this afternoon, if you please," said Uncle,
"only don't insist upon going too early ; we will
walk there about sunset."

"Now I think of it," added Uncle Hiram, "I have

another story of the sagacity of a dog who belonged
to one of my friends, and which is a pretty fair
match for Harry's Ranger. I wish he was here to
hear it.

"This was a large, powerful, quiet dog, of the
Newfoundland family, who was trained to go to
market, and do other errands for the family. They
would wrap the money in paper, put it into a bas-

THE PLAYMATE.

ket, and send Rover with it to the grocer's, or to
the market. The change, if any, was sent back in
the same way, with the articles purchased, and always were safely delivered.

"His master had a little boy who went to school,
and Rover was sometimes employed to carry the
boy's cap, or some other article. One day, when
it rained, Rover was called up, and told to carry

Charley's India rubbers to him, at school. He took
them in his mouth, and was going out of the door
with them, when he chanced to notice Charley's
cap, on the hat tree. He dropped the shoes, took
down the cap, and then for some time puzzled himself to get hold of the three articles together. At
length, stopping and taking an earnest look at the
matter, he very deliberately took up the shoes,
placed them in the cap, and then marched gravely
off, with an expression of perfect satisfaction that
was amusing to the lookers-on, as it was comfortable to himself."

THE POLAR BEAR.

THE POLAR BEAR.

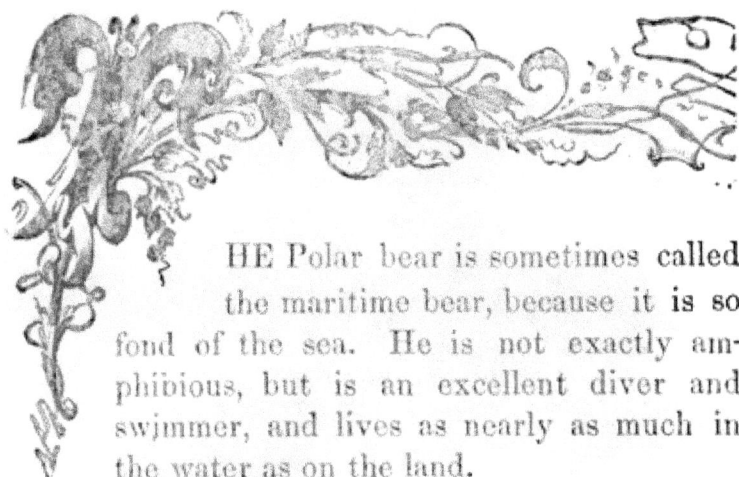

HE Polar bear is sometimes called the maritime bear, because it is so fond of the sea. He is not exactly amphibious, but is an excellent diver and swimmer, and lives as nearly as much in the water as on the land.

The Polar bear is always white, very large, powerful, ferocious, and daring ; a terrible fellow to encounter unless you are remarkably well armed. He is more fond of animal food than any other species of bear, though equally capable of living on vegetable food. He finds an ample supply even in the desolate regions where he chooses to reside, of seals, young whales, and the carcasses of whales, which are thrown out by the whalers, after they have taken what they want for oil, etc. How he manages to live in such regions of perpetual ice, it is difficult to imagine ; but he is never found except in the high northern latitudes, along the borders of ice-bound seas. He seems to require a large range of coast for his domain ; for he never comes down

into Siberia or Kamschatka, on the Eastern Continent, or to the same latitudes on the Western, except occasionally to the upper shores of Hudson's
Bay. He is not even found in the islands that lie
between the two continents. He is sometimes,
though very rarely, caught out of his latitude.
This is when some field of ice, on which he has fixed his temporary residence, breaks away from its
moorings, and is floated by the currents out into
the open sea. Some of them perish in this way,
not being able to regain the land, and their ice-
boat melting under them as it comes into a warmer
region. Some of them are taken or killed by the
sailors who discover them in this situation, though
it is generally found a very dangerous kind of sport
to meddle with them.

The Polar bear is very seldom seen in our caravans or menageries, because in the first place it is
almost impossible to catch them, and in the second
place, quite impossible to keep them alive in our
warm climate. There was one, and a very fine
large one, exhibited in New York in the spring of
1826. Though the weather was very cool at the
time, he suffered greatly, bathing himself in cold
water as often as he could, and seeming never
satisfied except when he could have ice in his cage
to live upon. Uncle Merry says he saw him, and
he was sweating and panting like a race-horse in
August.

THE NAMES OF ANIMALS.

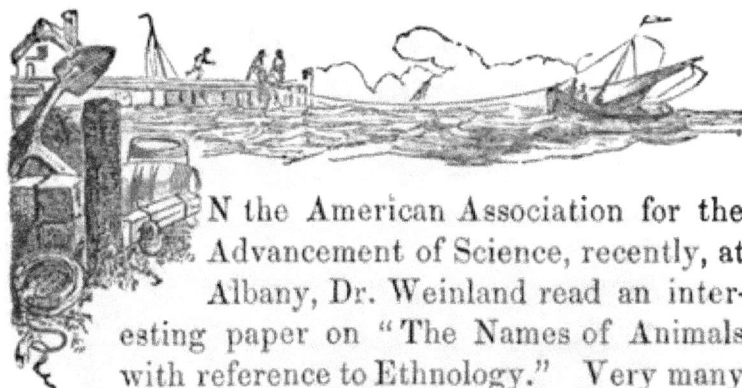

N the American Association for the Advancement of Science, recently, at Albany, Dr. Weinland read an interesting paper on "The Names of Animals with reference to Ethnology." Very many of the names of the North American animals are taken from European animals—thus, buffalo, grouse, robin, lizard, chamois. Nations have only names for their native animals. Thus, lion in all modern languages, is leo hardly changed.* The camel and the tiger derive their names from their native countries, other nations adopting these names with slight modifications. The elephant is so called in all countries. The ass got his name from the old Hebrews. The hare* and the deer, which occur both in Europe and Asia, and have two names, one native in each country—the former *lepus*, and the later *cervus*. Nations try to reduce all foreign animals to the names of their own, by adding a descriptive designation, as Guinea-pig, camel-leopard, river-horse, etc. The Anglo-Saxons who lived on the sea, had names for all sea-animals, but the Germans of the interior called them all by some land name, with the addition of " sea," thus, sea-horse, sea-dog, sea-lion, sea-tiger, sea-mouse, sea-devil.

Almost all animals were originally named from their qualities. The name of the ass comes from a root, meaning " walk slowly ;" the serpent to " glide quickly ;" the rabbit to " burrow in the ground." Prof. Haldeman said reindeer meant " running animal ; fox is from the Greek *phuxos*, " sharp ;" serpent from the Latin *serpo*, " to creep ;" and tiger from the Persian, " an arrow." Indian tribes call a lion by a name meaning " having a long tail ;" a horse by a name meaning " like a deer ;" a mole, " having his right hand on the left shoulder ;" a squirrel by a name meaning " he can stick fast in a tree." The Indians have also a name for a horse meaning " having only one toe." Apropos of names, it was remarked that the potato is called in German the " ground pear."

THE bound of the tiger, when springing upon his prey, is tremendous, extending, as it is said, to the distance of 100 feet. It is from this spring that the animal gets his name. He, as it were, " shoots himself at his prey ;" and tiger, in the Arminian language, signifies an *arrow*—the name also given to the river Tigris, on account of its velocity.

A SPORTING FISH.

A DEAD SHOT.

AN interesting account is given in the eleventh number of the *Edinburgh Philosophical Journal*, of the Jaculator fish of Java, by a gentleman who had an opportunity of examining some specimens of it in the possession of a chief.

The fish were placed in a small circular pond, from the centre of which projected a pole upward of two feet in height ; at the top of this pole were inserted several small pieces of wood, sharpened at the points, on each of which were transfixed some insects of the beetle tribe.

When all had become quiet, after the beetles had been secured, the fish, which had retired during the operation, came out of their hiding-places, and began to circle round the pond.

One of them at length rose to the surface of the water, and, after steadily fixing its eyes for some time upon an insect, discharged from its mouth a small quantity of water-like fluid, with such force and precision of aim, as to drive the beetle off the twig into the water, where it was instantly swallowed.

After this, another fish came and performed a similar feat, and was followed by the rest, till all the insects were devoured.

The writer observed, that if a fish failed in bringing down its prey at the first shot, it swam around the pond until it again came opposite the same object, and fired again.

In one instance, he remarked one of the fish returned three times to the attack, before it secured its prey ; but in general, they seemed to be very expert shots, bringing down the game at the very first discharge.

The Jaculator, in a state of nature, frequents the banks of rivers in search of food. When it spies a fly settling on the plants that grow in shallow water, it swims on to the distance of from five to six feet of them, and then with surprising dexterity, ejects from its tubular mouth a single drop of fluid, which rarely fails to strike the fly into the water, where it is immediately swallowed.

THINK OF IT.

A HUMMING-BIRD once met a butterfly, and, being pleased with the beauty of its person and the glory of its wings, made an offer of perpetual friendship.

"I cannot think of it," was the reply, "as you once spurned me, and called me a crawling dolt."

"Impossible," exclaimed the humming-bird, " I always entertained the highest respect for such beautiful creatures as you."

"Perhaps you do now," said the other ; "but when you insulted me, I was a caterpillar. So let me give you this piece of advice : never insult the humble, as they may one day become your superiors."

Boys and girls, think of this.

HUNTING DEER IN THE HIGHLANDS.

THE DEER.

THE hart is an animal of the antelope species; in size it is rather smaller than the fallow deer. Its colors vary somewhat in the different countries in which it is found. It is generally, however, of a dusky brown, mixed with red; the body underneath the breast and the inside of the limbs are white; but on the head, back, and outside of the limbs, the hair is considerably darker than on any other parts of the body. The orbits of the eyes are white, and there is a small patch of the same on each side of the forehead. The horns are perfectly black, and have three curves; they are marked with circles almost to the top; they are sixteen inches long. The female has no horns. This animal has a sort of substance in its stomach which is called bezoar, which is sometimes of a blood color, sometimes pale yellow, and of all the shades between the two. It is as hard as stone, and is generally glossy and smooth, with a smell which is considered very agreeable. It varies in size from that of an acorn to that of an egg ; and the larger the size, the more valuable it is. There was a time when a stone of this kind, weighing four ounces, sold in Europe for above two hundred

ANTELOPES.

pounds, but at present they are of comparatively little value. The word bezoar is derived from the Arabic language, where it signifies antidote, or counter poison. It has been given for various diseases, such as palpitation of the heart, colic, jaundice ; and in those countries where the price, and not the real use of the medicine is considered, it has been given for almost every disease which can be mentioned. It probably possesses merely the virtues of common chalk, and is only used where the knowledge of medicine has advanced but little.

The antelope is a very graceful creature, particularly when running. It is found principally in the hilly parts of the countries which it inhabits. It is very cunning, and requires careful watching and much tact to be shot or taken. They run in herds, and rarely lie down altogether; but by an instinct given them by Providence, some are always on the watch, and when they are fatigued, they give notice to those who have rested, who arise at once and relieve the sentinels of the preceding hours, and thus they often preserve themselves from the attacks of wolves and huntsmen. They are exceedingly swift, and will outrun the fleetest horse or greyhound. If, however, they are bitten by a dog, they at once fall down, nor will they even offer to rise again.

Deer hunting is a favorite amusement in many countries. These animals are found in America, from Canada in the North, to the banks of the Oronoco in South America; also in many parts of England; although they are not so plentiful as formerly, they are, however, still to be found in the Highlands of Scotland in considerable numbers.

A MONKEY'S MEMORY.

A KNOWING MONKEY.

AUTHORS generally think that the monkey race are not capable of retaining lasting impressions, but their memory is remarkably tenacious when striking events call it into action. A monkey which was permitted to run free, had frequently seen the men-servants in the great country kitchen, with its huge fire-place, take down a powder-horn that stood on the chimney-piece, and throw a few

grains into the fire, to make Jemima and the rest
of the maids jump and scream, which they always
did on such occasions very prettily. Pug watched
his opportunity, and when all was still, and he had
the kitchen all to himself, he clambered up, got possession of the well-filled powder-horn, perched himself very gingerly on one of the horizontal wheels
placed for the support of saucepans, right over the
warming ashes of an almost extinct wood-fire,
screwed off the top of the horn, and reversed it
over the grate. The explosion sent him half way
up the chimney. Before he was blown up, he was
a snug, trim, well-conditioned monkey as ever you
would wish to see on a summer day ; he came down
a carbonated nigger in miniature, in an avalanche
of burning soot. The weight with which he pitch-
ed upon the hot ashes, in the midst of the general
flare up, aroused him to a sense of his condition.
He was missed for. days. Hunger at last drove
him forth, and he sneaked into the house, close-sin-
ged, begrimmed, and looked scared and ugly. He
recovered with care ; but, like some great person-
ages, he never got over the sudden elevation and
fall, but became a sadder if not a wiser monkey.
If ever Pug forgot himself and was troublesome,
you had only to take down a powder horn in his
presence, and he was off to his hole like a shot,
screaming and shattering his jaws like a pair of cas-
tanets.

Monkey are *quadrumana*—four-hand. But, while four feet contribute to swiftness, and four hands to agility, that combination of the two, as in man, which gives two hands and two feet, with separate and distinct functions, is not only far the most convenient, but confers far greater power, variety, and versatility of action. To no animal, except man, is the upright position natural. The monkey assumes it occasionally, for convenience, or in obedience to the training of a human master.

There are three distinct families of monkeys, differing from each other in some respects widely, but having the same general characteristics.

The SIMIADÆ include all the animals of the Old World, known as apes, monkeys, and baboons. The ape has no tail, the monkey a long one, and the baboon a short one.

The *Chimpansé* is a species of ape, approaching more nearly to man than any other animal. Even in a natural state he sometimes walks erect, supporting himself with a cane. Some of them have been tamed and trained to various kinds of useful labor, such as bringing water from the well, washing dishes, and even waiting upon table. It is a native of Central Africa.

The Ourang-Outang belongs also to the ape family. His countenance resembles the human face more than that of any other. His dwelling is principally in trees, and he moves with difficulty on the ground. He is of a quiet, grave, and even melancholy disposition. He has great strength, and when excited-to rage is often very savage. He belongs chiefly to the peninsulas and islands of Eastern Asia.

The Ourang-Outang, which, in the Malay language, means "*wild man*," is incapable of walking upright. He is not very large, being about two feet seven inches high. The hair on his back is five or six inches long.

The *Baboon* has usually a very short tail, or none at all. It is distinguished from the ape and the monkey by the protuberance of the muzzle, which gives it a ferocious aspect. It has a loud and discordant voice, and is less companionable and docile than the other species. It is revengeful, and retains for a long time a remembrance of an injury done it.

The *Monkey*, properly so called, is also of the ape species—a bright, smart, mischievous, cunning fellow, making lots of fun for children, in all our towns and cities, but often very cruelly treated by their masters.

We should not take as much pleasure in witnessing the curious antics of monkeys, if we knew how hardly, and under what severe treatment, they learned their lessons.

In England, a fight was instigated between a monkey and bull-dog, on a wager of three guineas

to one, that the dog would kill the monkey in six
minutes. The owner of the dog agreed to permit
the monkey to use a stick about a foot long. Hundreds of spectators assembled to witness this in-
human sport. The owner of the monkey taking
from his pocket a thick, round rule, about a foot
long, threw it into the hand of the monkey, saying,
"Now look sharp—mind that dog." "Then here
goes for your monkey," cried the butcher, letting
the dog loose, which flew with a tiger-like fierceness at him. The monkey, with astonishing agility,
sprang at least a yard high, and falling on the dog,
laid fast hold to the back of his neck with his teeth,
seizing one ear with his left paw, so as to prevent
his turning to bite. In this unexpected situation,
Jack fell to work with his rule upon the head of
the dog, which he beat so forcibly and rapidly,
that the creature cried out most eloquently. In a
short time the dog was carried off in nearly a lifeless state, with his skull fractured. The monkey
was of the middle size.

THE LEOPARD.

HAVE you ever seen a wild leopard? Probably not, and very probably you would not like to see him very near, unless he was chained or caged. His appearance in the wild state is exceedingly beautiful, his motions in the highest degree easy and graceful, and his agility in bounding among the rocks and woods quite amazing. He usually shuns a conflict with a man, but, when driven to desperation, becomes truly a formidable antagonist.

Two African farmers, returning from hunting the hartebeest (*antilope babulis*), roused a leopard in a mountain ravine, and immediately gave chase to him. The leopard at first endeavored to escape by clambering up a precipice ; but being hotly pressed, and wounded by a musket ball, he turned upon his pursuers with that frantic ferocity peculiar to this animal on such emergencies, and springing on the man who had fired at him, tore him from his horse to the ground, biting him at the same time on the shoulder, and tearing one of his cheeks severely with his claws. The other hunter seeing the danger of his comrade, sprang from his horse, and attempted to shoot the leopard through the head ; but, whether owing to trepidation, or the fear of wounding his friend, or the quick motions of the animal, he unfortunately missed. The leopard, abandoning his prostrate enemy, darted with redoubled fury upon his second antagonist, and so fierce and sudden was his onset, that before the boor could stab him with his hunting-knife, the savage beast struck him on the head with his claws, and actually tore the scalp over his eyes. In this frightful condition, the hunter grappled with the leopard : and, struggling for life, they rolled together down a deep declivity. All this passed far more rapidly than it can be described in words. Before the man who had been first attacked could start to his feet and seize his gun, they were rolling one over the other down the bank. In a minute or two he had reloaded his gun, and rushed forward to save the life of his friend. But it was too late.

The leopard had seized the unfortunate man by the throat, and mangled him so dreadfully, that death was inevitable ; and his comrade (himself severely wounded) had only the melancholy satisfaction of completing the destruction of the savage beast, already exhausted with the loss of blood from several deep wounds by the desperate knife of the expiring huntsman.

The fur of the leopard (*leo-pard*, or spotted lion) is yellow, with ten ranges of black spots, or clusters of spots, on each side.   Each spot is made up of a number of smaller spots.

THE SHEPHERD'S DOG.

A GENTLEMAN sold a considerable flock of sheep to a dealer, which the latter had not hands to drive. The seller, however, told him he had a very intelligent dog, which he would send to assist him to a place about thirty miles off; and that when he reached the end of his journey, he had only to feed the dog and desire him to go home. The dog accordingly received his orders, and set off with the flock and the drover; but he was absent for so many days that his master began to have serious alarms about him, when one morning, to his great surprise, he found his dog returned with a very large flock of sheep, including the whole that he had lately sold. The fact turned out to be, that the drover was so pleased with the colley that he resolved to steal him, and locked him up till the time when he was to leave the country. The dog grew sulky, and made various attempts to escape, and one evening he succeeded. Whether the brute had discovered the drover's intention, and supposed that the sheep were also stolen, it is difficult to say; but by his conduct it looked so, for he immediately went to the field, collected the sheep, and drove them all back to his master.

PECULIARITIES OF THE REINDEER.

THE REINDEER.

THE reindeer is the color of the stag, and is not much larger. The horns of this animal are somewhat higher than those of the stag, but more crooked, hairy, and not so well furnished with branches. Of the milk of the females they make good butter and cheese. These animals, indeed, constitute the greatest and almost the only riches of the Fin Laplanders. In Finmark there are vast numbers of them, both wild and tame, and many a man there has from six or eight hundred to a thousand of these useful creatures, which never come under cover. They follow him wherever he is pleased to ramble, and when they are put to a sledge, transport his goods from one place to another. They provide for themselves, and live chiefly on moss, and on the buds of leaves and trees. They support themselves on very little nourishment, and are neat, and clean, and entertaining creatures. It is remarkable when the reindeer sheds his horns, and others rise in their stead ; they appear at first covered with a skin, and till they are of a finger's length, are so soft that they may be cut with a knife like a sausage, and are delicate eating, even raw ; therefore the huntsmen, when far out in the country, and pinched for the want of food, eat them, and find that they satisfy both their hunger and their thirst. When the horn grows bigger, there breeds within the skin a worm which eats away the root. The reindeer has over his eyelids a kind of skin, through which he peeps, when otherwise, in the hard snows, he would be obliged to shut his eyes entirely—a singular instance of the benevolence of the great Creator in providing for the wants of each creature according to its destined manner of living.

THE COACH DOG.

THIS dog is a native of Dalmatia, a mountainous
district of European Turkey. He has been
domesticated in Italy for upwards of two centuries,
and is now often to be met with both in Europe
and this country.

The Dalmatian is often used as a pointer, to
which his natural propensity more inclines him.
He is handsome in shape ; his general color is white,
and his whole body and legs are covered with small
irregular-sized black or reddish-brown spots.   A
singular opinion prevailed at one time in England,
that this beautiful dog was rendered more hand-

some by having his ears cropped : this custom has now gone out of use.

The chief use of this dog seems to be as an attendant upon a carriage, for which the symmetry of his form and beauty of his skin peculiarly fit him.

My young readers may have before met with the following remarkable instance of sagacity in a dog : " A surgeon, of Leeds, in England, walking in the suburbs of that town, found a little spaniel, who had been lamed. This dog, which probably had its name from Spain, belongs to a differing species from the coach-dog. Well, the surgeon carried the poor little lame animal home, bandaged up his leg, and, after two or three days, turned him out. The dog returned to the surgeon's house every morning, till his leg was perfectly well.

At the end of several months, the spaniel again presented himself, in company with another dog, who had been lamed ; and he intimated, as well as piteous and intelligent looks could intimate, that he desired the same kind assistance to be rendered to his friend, as had been bestowed upon himself."

The fame of an English dog has been deservedly transmitted to posterity by a monument in basso relievo, which still remains on the chimney-piece of the grand hall, at the Castle of Montargis, in France. The sculpture, which represents a dog fighting with a champion, is explained by the following :

Aubri de Mondidier, a gentleman of family and fortune, traveling alone through the Forest of Bondi, was murdered and buried under a tree. His dog, an English bloodhound, would not quit his master's grave for several days ; till at length, compelled by hunger, he proceeded to the house of an intimate friend of the unfortunate Aubri's at Paris, and by his melancholy howling seemed desirous of expressing the loss they had both sustained. He repeated his cries, ran to the door, looked back to see if any one followed him, returned to his master's friend, pulled him by the sleeve, and with dumb eloquence entreated him to go with him.

The singularity of all these actions of the dog, added to the circumstance of his coming there without his master, prompted the company to follow the animal, who conducted them to a tree, where he renewed his howl, scratching the earth with his feet, and significantly entreating them to search that particular spot. On digging, the body of the unhappy Aubri was found.

Some time after, the dog accidently met the assassin, who is styled, by all the historians that relate this fact, the Chevalier Macaire ; when instantly seizing him by the throat, he was with great difficulty compelled to quit his prey. In short, whenever the dog saw the chevalier, he continued to pursue and attack him with equal fury. Such obstinate virulence in the animal, confined only to

Macaire, appeared very extraordinary ; especially to those who at once recollected the dog's remarkable attachment to his master, and several instances in which Macaire's envy and hatred to Aubri had been conspicuous.

Additional circumstances created suspicion ; and at length the affair reached the royal ear. The king (Louis VIII) accordingly sent for the dog, who appeared extremely gentle till he perceived Macaire in the midst of several noblemen, when he ran fiercely toward him, growling at and attacking him as usual.

The king, struck with such a collection of circumstantial evidence against Macaire, determined to refer the decision to the chance of battle ; in other words, he gave orders for a combat between the chevalier and the dog. The lists were appointed in the Isle of Notre Dame, then an uninclosed, uninhabited place, and Macaire was allowed for his weapon a great cudgel.

An empty cask was given to the dog as a place of retreat, to enable him to recover breath. Every thing being prepared, the dog no sooner found himself at liberty, than he ran round his adversary, avoiding his blows, and menacing him on every side, till his strength was exhausted ; then, springing forward, he griped him by the throat, threw him on the ground, and obliged him to confess his guilt, in the presence of the king and his whole

court. In consequence of this, the chevalier, after
a few days, was convicted upon his own acknowledgment, and beheaded on a scaffold in the Isle of
Notre Dame.

The Newfoundland dog, in a state of purity, and
uncontaminated by a mixture of an inferior race, is
certainly the noblest of the canine tribe. His great
size and strength, and majestic look, convey to the
mind a sort of awe, if not fear, but which is quickly
dispelled when we examine the placid serenity and
the mild expressive intelligence of his countenance,
showing at once that ferocity is no part of his disposition.

The full-sized Newfoundland dog, from the nose
to the end of the tail, measures about six feet and
a half, the length of the tail being two feet. This
dog was but recently introduced into Europe from
the island whose name he bears, and may be considered as a distinct race.

The Newfoundland dog is docile to a very great
degree, and nothing can exceed his affection. Naturally athletic and active, he is ever eager to be
employed, and seems delighted to perform any
little office required of him. Nature has given him
a great share of emulation, and hence to be sur-
passed or overcome is to him the occasion of great
pain. Active on every emergency, he is the friend
of all, and is naturally without the least disposition
to quarrel with other animals. He seldom or never

offers offence, but will not receive an insult or injury with impunity. Such is the capacity of his
understanding, that he can be taught almost every
thing which man can inculcate, and of which his
own strength and frame are capable. His sagacity
can only be exceeded by his energies, and he per-
severes with unabated ardor in whatever shape he
is employed, and while he has a hope of success he
will never slacken in his efforts to attain it. The
amazing pliability of his temper peculiarly fits him
for the use of man, and he never shrinks from any
service which may be required of him, but under-
takes it with an ardor proportionate to the difficulty
of its execution. Taking a singular pride in being
employed, he will carry a stick, a basket, or a bun-
dle, for miles, in his mouth, and to deprive him of
any of these is more than a stranger could accom-
plish with safety.

Sagacity and a peculiar faithful attachment to
the human species are characteristics inseparable
from this dog, and hence he is ever on the alert to
ward off from his master every impending danger,
and to free him from every peril to which he may
be exposed. He is endowed with an astonishing
degree of courage, whether to resent an insult or
to defend his friends, even at the risk of his own
life.

The qualifications of this dog are extensive indeed ; as a keeper or defender of the house, he is

far more intelligent, more powerful, and more to be
depended upon than the mastiff.   As a watch dog,
and for his services upon navigable rivers, none can
compete with him ; and various sportsmen have
introduced him into the field as a pointer with great
success, his kind disposition and sagacity rendering
his training an easy task.

, The usual fate of other fine dogs attends this
generous race among us ; they are too often de-
graded and degenerated by inferior crosses, which
with so noble an animal should be avoided by every
possible means.

At the commencement of an action which took
place between the Nymph and Cleopatra, during
the late war, there was a large Newfoundland dog
on board the former vessel, which the moment the
firing began ran from below deck, in spite of the
endeavors of the men to keep him down, and climbing up into the main-chains he there kept up a
continual barking, and exhibited the most violent
rage during the whole of the engagement.

When the Cleopatra struck, he was among the
foremost to board her, and there walked up and
down the decks, seemingly conscious of the victory
he had gained.

THE RHINOCEROS.

OF all South African animals, not the least curious, perhaps, is the rhinosceros. He is, moreover, an inhabitant of Bengal, Siam, China, Java and Ceylon ; but these are a different species from those found in Africa. Thus, there are the black and the white, and both species are extremely fierce, and excepting the buffalo, are the most dangerous of all animals in Southern Africa. His appearance is not unlike an immense hog shorn of his bristles, except a tuft at the ears and tail. As if in mockery of its great size, its eyes are ludicrously small, so as to be almost imperceptible.

"Two officers belonging to the troops stationed at Dunapore, went down to the river to shoot and hunt, and they had heard at Derrzapore of a rhinoceros having attacked and murdered travelers in this region. One day, before sunrise, as they were about starting out to hunt, they heard a violent uproar, and on looking out, found that a rhinoceros was goring their horses, both of which, being fastened by head and heel, were unable to resist or escape. Their servants took to their heels, and concealed themselves in a neighboring jungle. The gentlemen had just time to climb up into a small tree close by, before the furious beast, having devoured the horses, turned his attention to the masters. They were barely out of his reach ; so after keeping them for some time in terrible suspense, vainly endeavoring to dislodge them, seeing the sun rise, he retreated, not, however, without glancing back occasionally, as if regretting the loss of so fine a feast."

"Once," says Mr. Oswell, "as I was returning from an elephant chase, I observed a huge rhinoceros a short distance ahead. I was riding a most excellent hunter, the best and fleetest steed I possessed during my shooting excursions in Africa ; but it was a rule with me never to pursue a rhinoceros on horseback, for this reason, that they were more easily surprised and killed on foot. On this occasion it seemed as if fate had interfered. Turning to my servant, I called out, 'That fellow has a magnificent horn ; I *must* have a shot at him !'

"Saying this, I clapped spurs to my horse, who

soon brought me alongside the huge beast, and the
next instant I had lodged a ball in his body—but as
it turned out, without effect. On receiving my
shot, the rhinoceros, to my surprise, instead of retreating, stopped short, turned round, and having
eyed me for some seconds, walked toward me. I
never dreamed of danger, but instinctively turned
my horse's head. It was too late, for although the
rhinoceros had been only *walking*, the distance was
so inconsiderable that contact was unávoidable. In
a moment I saw his head bend low ; with a thrust
upward he struck his horn into the ribs of the horse
with such force as to penetrate to the saddle on the
opposite side, where its sharp point pierced my leg.
The violence of the blow was so tremendous as to
cause the horse to perform a complete somerset in
the air, coming down heavily on his back. As for
myself, I was violently precipitated to the ground.

"The rhinoceros seemed satisfied with his revenge, and started off on a canter. My servant
having now come up, I rushed up to him, almost
pulled him from his horse, leapt into the saddle and
without a hat, my face streaming with blood, pursued the retreating animal, and had soon the satisfaction of seeing him fall lifeless at my feet. My
friend, by whom I was accompanied on this journey,
soon after joined me, and seeing my head and face
covered with blood, supposed me to be mortally
wounded ; but, with the exception of a severe blow

on the head, caused by the iron stirrups, I received no injury, although my much prized horse was killed on the spot.

"On another occasion, while wending my steps toward my camp on foot, I espied at no great distance two rhinoceroses—called keitloa. They were feeding, and slowly approaching me. I immediately crouched, and quietly awaited their arrival ; but

THE TWO HORNED RHINOCEROS.

though they soon came within range, I was unable to fire, as they were facing me, and a shot in the head is useless. In a short time they had approached so close that owing to the level open nature of the ground, I could neither retreat nor advance, and my situation was most critical. I was afraid to fire, for even had I succeeded in killing one, the other would in all likelihood have run over and trampled me to death. In this dilemma, the thought struck me that on account of their bad sight I might possibly save myself by running past them. No time was to be lost, and as the foremost animal almost touched me, I stood up and dashed past it. The brute, however, was too quick for me, and before I had gone many steps, I heard a violent snorting at my heels. I had only time to fire my gun at random toward him, when I felt myself impaled on his horn. The shock completely stunned me. The first return to consciousness was, I recollect, finding myself seated on one of my ponies, and a Caffre leading it. I had an indistinct idea of having been hunting, and on seeing the man, asked why he did not follow the animal. By accident I touched my right hip, and on looking at my hand, found it clotted with blood. While in my confused state, trying to understand what it meant, I saw my men coming toward me, who told me they were coming to fetch my body, as they had been told I was killed. The wound I received was dangerous, and though after a long time it healed, still the scars will remain as long as I live."

THE CROCKODILE.

ANECDOTES OF THE CROCODILE FAMILY.

THE crocodiles of the eastern continent and the alligators of our own are all of one family, but there appears to be some difference in their dispositions. Darwin describes one of them as follows :—

" Erewhile emerging from the brooding sand,
 With tiger paw he prints the brineless strand ;
 High on the flood with speckled bosom swims,
 Helmed with broad tail, and oared with giant limbs ;
 Rolls his fierce eyeballs, clasps his iron claws,
 And champs with knashing teeth his massy jaws :
 Old Nilus sighs through all his cane-crowned shores,
 And swarthy Memphis trembles and adores."

This may stand as a good and fair likeness of the whole—a sort of family portrait of the grandpapa of crocodiles and alligators. Thus the governor of Angostura informed Mr. Waterton, that while he was one fine evening walking by the banks of the Oronoco, he saw a large cayman rush out of the river, seize upon a man, and carry him away in his horrid jaws. "The screams of the poor fellow were terrible, as the cayman was running off with him.

He plunged into the river with his prey ; we instantly lost sight of him, and never saw or heard him more."

So also in regard to the African species, we may, among many other recorded examples of their fierceness, recall to mind the circumstance of Mungo Park's negro guide Isaaco being twice seized by a crocodile while crossing the Ba Woolima with his asses, and escaping immediate death only by his presence of mind enabling him to gouge the eyes of the insatiate monster with his thumbs. He gained the shore bleeding profusely with a deep wound in each thigh, and the marks of several teeth upon his back. He was unable to renew the journey for six days.

These, and many other anecdotes of the same kind, justify the general bad character given to these creatures. But, on the other hand, it appears that some of them are of a more gentle nature. It is perfectly well known that the ancient Egyptians worshiped crocodiles, and it appears that the priests kept many of them, which seem to have been quite tame and friendly. So Mr. Audubon informs us, that in North America the alligators, in some parts, are so disinclined to annoy the human race, that he and his companions have often waded up to the waist among hundreds of them, while the cattle-drivers may be seen beating them away with staves, before they cross the rivers with their live stock ;

for it is admitted that they readily attack cattle, and will seize upon such animals as dogs and deer, or even horses.

Although a full grown, case-hardened crocodile, with its armature of "scaly rind," and formidable jaws beset with bristling teeth, need fear nothing short of a rifle-bullet through the eye, or a volley of slugs in the softer part of the abdomen, yet the eggs and young fall a frequent prey to many natural enemies. The ichneumons of Egypt, the otters and even ibises of the new world, and the great tortoises belonging to the genus *Trionyx*, attack them generally in one or other of these defenceless states ; while, at least so far as concerns the North American species, the male parent, repudiating all the claims of filial affection, throws, not his arms, but jaws around his unprotected young, and gulps them down in dozens. However, the negroes will attack even the adult animals, and kill them by separating the tail from the body by blows of their hatchets. The oil (obtained by boiling) is used for machinery ; and a practice prevailed, at one time, of making boots and shoes of alligator leather. The South American Indians eat the tail of these creatures, and they catch the owner of the tail by means of a small hook baited with a bird, or any small quadruped, and fastened to a tree by an iron chain. "The flesh," according to Catesby, "is delicately white, but hath so perfumed a taste and

smell that I never could relish it with pleasure."
The eggs of the crocodile are regarded as a luxury
by some of the African tribes.

In regard to the geographical distribution of
these great reptiles, we need scarcely inform our
readers that no species occurs in Europe. Neither
has any been found in New Holland. The caymans
or alligators are peculiar to America; the crocodiles, properly so called, are natives of both the
old and new world; the gavials are confined to
Asia, to the verdent banks.

GRATIAS THE CATERPILLAR.

GRATIAS had several queer adventures. Once
a large green frog, with a cold nose and gog-
gle eyes, snapped at him as he was looking over
the edge of the fish-pond ; but there was some slimy
moss on the stone where Freckle stood ; and just
as his mouth was about to close on Gratias, his long
hind legs slipped and sprawled ; he went back into

the water with a splash, and our brown friend
traveled off so fast, he never saw the garden wall
before him till he bumped his head against it.
Then, once, he had gone to sleep in the very middle of a red rose—the last one on the bush, for it
was now autumn—and the rose being picked by a
very little white hand, that belonged to Miss Sac-
charissa, Gratias began to quirl for joy ; he thought
she would be good to him if he was not pretty, for
her blue eyes were so very soft and shallow, just
like the pond on a summer day ; but when Saccha-
rissa saw the innocent worm, she gave a loud shriek
and threw rose and all on to the gravel path so hard,
that Gratias had scarcely time to make a ball of
himself and roll away, to hide his bruised head and
his hurt feelings behind the garden roller for two
days. But a diet of chickweed and rain water
cured both those ailments, and soon he crept out
again over the big roller, which just at that hour
the gardener was accustomed to use, and poor
Gratias began to feel it move under him before he
was half way across it, and expected nothing less
than to be directly crushed to a jelly ; but the
gardener found his roller was out of order, a loose
screw threatened to let the handle go every moment ; and while he replaced that, Gratias had
time to save himself, and dropping to the ground,
toddled away, half a mind to be discouraged and
say he would not try to live any more, he was so

lonely and so ugly, and so full of fear ; however, a
little honey-bee just then began to sing on a late
bean-flower, and her song was so gay and so good,
that the worm found himself trying to sing too.

> " Buzz, buzz, buzz away !
>> Making honey
>> When it's sunny,
> Sleeping all the rainy day.

> " Buzz, buzz, busy bee,
>> All the posies
>> Are not roses,
> But they all are sweet to me.
>> Buzz, buzz away !"

"Whew !" said Mr. Powsy, who turned the corner just then ; "a nice little song, Mrs. Sweeting !
do you think winter won't come ?"

"I shall go to sleep then, sir, and there's honey
in the hive," answered the little woman.

But Gratias shivered. " Is it almost winter, Mr.
Powsy ?"

" Yes, creeper crawler ; almost time for the white
frosts. I've been hard at work, to-day, picking out
a place for my hole ; soon I shall have to dig it."

" And where are all the creatures I know going
this year ?" said Gratias, in a dismal tone.

"Oh ! I go to sleep. Buzz and Mrs. Sweetser
stay and nod in their combs, Mrs. Pelopidan went
South yesterday, and Mrs. Roberts has taken a
house for the winter in the great barn ; when it is

very cold she may go to Maryland, I can't say. As
for Whiz, Fiz's brother, nobody seems to know exactly what **he will do.** *I* think he will die off.
Freckle, the frog, **is a low creature ;** he lives in the
mud, and comes out in the spring with such a host
of little polliwogs ! It is so absurd to have **children**
with tails, and no legs ! I don't see how he can be
so proud of the little wretches !"

" I wonder what **I shall do ?"** said Gratias ; but
Mr. Powsy **had hopped off** after a blue-bottle fly, so
he got **no answer. Then** he went up the nearest
tree and **lay in the sunshine, till** he felt so lazy and
dreamy that he **thought he would** spin a little ; and
he drew out **a nice** fine thread, longer than ever he
could before, till he thought how nice it would be
to spin himself a house for the winter, and resolved
to begin immediately ; so first he spun a stout **cord**
from the tree bough, and then a filmy veil **large**
enough for the outside of his house, an**d then** another and another layer, till he had b**ut** just room
to coil himself up and go to sleep, rocked by the
winds that began to blow cold and loud in the tree-
tops. But as he was getting very sleepy indeed,
he happened **to think** that he was **so fast** shut up
in his house that he **could not** possibly get anything to eat or drink, and what should he do ? For
a few minutes he was somewhat troubled, and
would have liked to unspin his new covering ; but
then he remembered that he had all his life been

taken care of, when he could not help himself, and he would not be afraid now ; so he curled down again, safe because he was helpless, and went sound asleep.

THE CHRYSALIS.

Now came the dim shape once more that Mrs. Pelopidan had seen, and took its stand by the grey house of the sleeping worm, to defend it from harm till spring should come. Gentian, the blue jay, that lived hard by, peered curiously at the swinging shell, but dared not touch it, for he saw the

awful shadow that stretched upward to the pure
skies, and kept guard over earth.

Flisk, the squirrel, chattered at a yard's distance,
about this queer nut to his wife Flisky, but came
no nearer ; and even the snow and rain beat to one
side, rather than freeze or wet the quiet home of
the hidden caterpillar.

At last spring came ; the grass began to shoot up
in the level meadows ; all the birds came back with
songs of pure love and joy ; the little wood-flowers
opened their soft eyes, and kissed the south wind
back again till it was as sweet as their own hearts ;
the tender rain wept for gladness, till all the buds
on the dim trees opened into leaves under its gentle caress ; and far and wide the grey woods melt-
ed into pale green masses ; the hill-sides grew
opal-colored with maple blossoms and bursting
buds ; the orchards blushed like rosy clouds on the
distant mountain slopes ; and all the world was so
happy, that a little stir of its new life came to Gratias where he slept, and the dim shape vanished in
the east. Warmer and warmer shone the sun on
the grey house, and the worm felt its glow through
every little bone ; he stretched himself well, and
the bands that seemed to hold him tightly, parted
gently ; he saw a tiny gleam of day and crept towards it, every motion growing easier and making
the spot of light wider, till at length he stood on
the outside of his winter dwelling in the noon-day

sun, dazzled and happy, but feeling as if he could not crawl.

"Whew!" said a well-known voice, and looking down he saw Mr. Powsy under the tree; "are you paid now for your patience, friend? Do you like your wings as well as Fiz did his?"

"Wings! have I got wings?" said Gratias.

"To be sure you have; sail across the pond and see yourself."

He spread the silken sails that now he felt on either side; lifted his dainty feet from the bough, and aided by a little puff of wind, away he glided with the most beautiful motion over flower-beds and paths to the great ponds, and poising above the blue surface, he looked down and saw himself—his ugly body was gone; his wings were gold-colored, all spotted with black and blue; his breast, mixed rings of black and gold; his eyes as bright as dew, and two slender, graceful, curling horns on either side of his head. He had not been so trustful and patient in vain; he was no more a worm, but a gay and beautiful butterfly, and he soared back to Mr. Powsy, almost too happy to fly straight.

"Ho! ho!" said the toad. "Now you're fine and must eat honey; I can't eat you now, if I wanted to. You must have a new name, friend; Gratias did very well for the worm, but the butterfly shall be called Gloria!"

And that was his name always.

THEDA'S PUSSY.

Is this *you*, my pussy ?
　Why, just now **I saw**
Your back rounded upward,
　And *nails* on each claw.

You were spitting *so* fiercely,
　Because little **Trip**
Would, in your nice breakfast,
　His saucy mouth dip.

'Twas an ungallant **action**
　In the dog, I own ;
But your cat indignation
　Was too roughly shown.

It is very low manners,
　To bluster and scratch ;
And it's worse, because useless—
　For Trip you're no match.

This is far more becoming—
  The soft velvet paw,
Which o'er cheek and o'er eyelid
  I now love to draw.

Run, and set your ball rolling;
  The ball you may strike—
Whiz it off to the corner,
  As hard as you like.

LEARNING TO ROLL BALL.

Now your lovely, my pussy,
  And mother smiles too;
Oh! we both think so pretty,
  The spry tricks you do

TIGER-HUNT WITH ELEPHANTS.

ELEPHANTS, HOW TAKEN AND MANAGED.

ARRIAN, a Greek writer of the second century, thus describes the ancient mode of catching elephants :—A large circular ditch is first made, inclosing space sufficient for the encampment of an army. The earth thus removed is heaped up on each margin of the ditch, and serves as a wall. In these walls there is one opening toward the south, with a bridge across the ditch, covered with earth and grass. In the outer wall are several excavations, near the bridge, in which the hunters secrete themselves, and watch, through loop-holes, the movements of the elephants. Several tame female elephants are placed in the inclosure, to attract the wild ones from without. When a sufficient number have entered the trap, the hunters issue from their hiding places, and take up the bridge. After a day or two, when the captives are somewhat weakened by want of food and water, they muster a large company of men with tame elephants, replace the bridge, and send the tame elephants into the inclosure. A battle ensues, which naturally terminates in favor of the tame animals, their opponents being quite exhausted by what they have previously suffered. The men now coming up tie their feet. After this, the process of taming and training them is not difficult.

It is remarkable, that in every mode of capturing the wild elephant, man avails himself of the docility of those he has already subdued. Birds may be taught to assist in insnaring other birds, but this is simply an effect of habit and training. The elephant, on the contrary, has an evident **desire to** join its master in subduing its own race. **It enters** into it with alacrity, and exercises ingenuity, courage, and perseverance, that are astonishing.

It is often noticed that large male **elephants, the** very ones that would be selected from **a** flock as most desirable for use, or for sale, are wandering **away by** themselves, apart from the herd. These **are** watched, and followed cautiously by day and night, with several trained females, called Koomkies. Approaching gradually nearer, and grazing with apparent indifference, the Koomkies at length press round their victim, and begin to caress him. If he is in good humor, and submits to **their** caresses, his capture is certain. The hunters cautiously creep under him, and while he is **dallying** with his new-found friend, bind his forelegs **together with a strong rope.** Some of the more wily of the **Koomkies** will not only protect their **masters,** while doing this, but actually assist in fastening the cords. Sometimes the hind legs are fastened in the same manner ; when the hunters retire to a distance to watch the motions of the captive. The Koomkies, satisfied that **he is** secure, **now** leave him. He at-

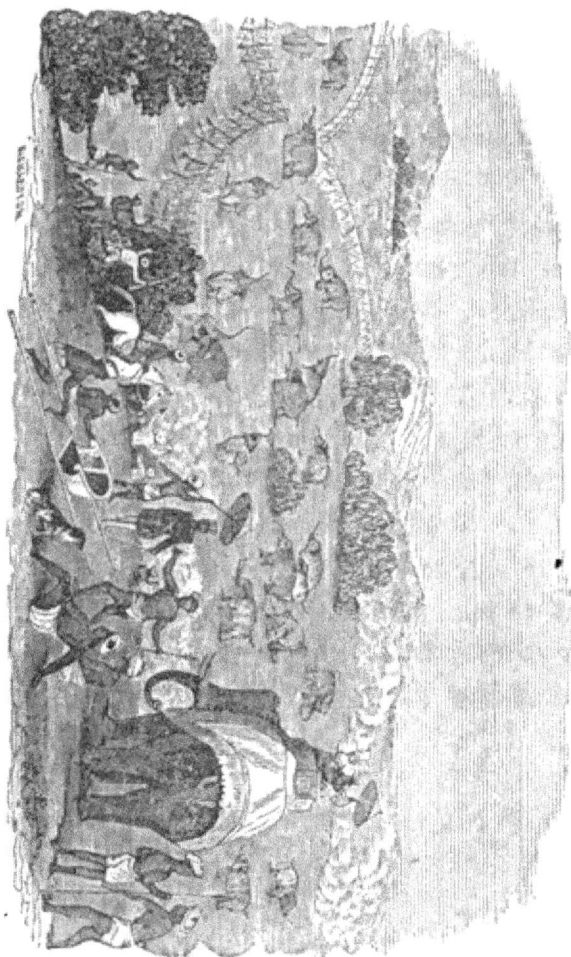

TRAPPING ELEPHANTS.

tempts to follow, but is unable. He now becomes
furious, throwing himself down and tearing the
earth with his tusks. If he succeed in breaking
the cords, and escaping to the forest, the trappers
dare not pursue him. If not, he is soon exhausted
with his own rage. He is then left until hunger
makes him submissive, when under the escort of
his treacherous friends, he is conducted to an inclosure, where he is fed, trained, and completely
subdued.

The inclosure, surrounded by a ditch, is still in
use in India. But not content with enticing their
victims to the place, they gather in large numbers,
and with fire-arms, and all kinds of noisy instru-
ments, drive whole herds of them in, the way being first strewed with the fruits they most like, to
tempt them onward. From this inclosure they
never come out till they are perfectly tamed. Each
elephant has his own *mahout*, or master, and will
obey no other.

THE FOOLISH MOUSE.

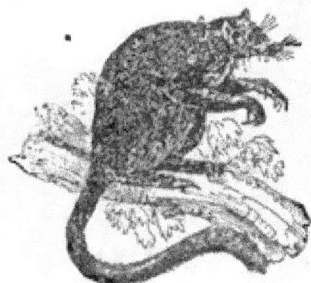

ALWAYS nibbling. little mouse,
  Fear you not your teeth to spoil,
Gnawing wood, cake, cheese, and nut-shells?
  Have you dentists with gold foil?

Betsey daily tries to kill you,
  Know you that, you silly elf?
Sure as fate, and will you—nill you,
  Springing trap is on that shelf.

There you go, pell-mell, head foremost,
  Anywhere you'll go for cheese—
Snap! now Betsey's trap has got you—
  That must be "von too much" squeeze.

MEN VS. ANIMALS.

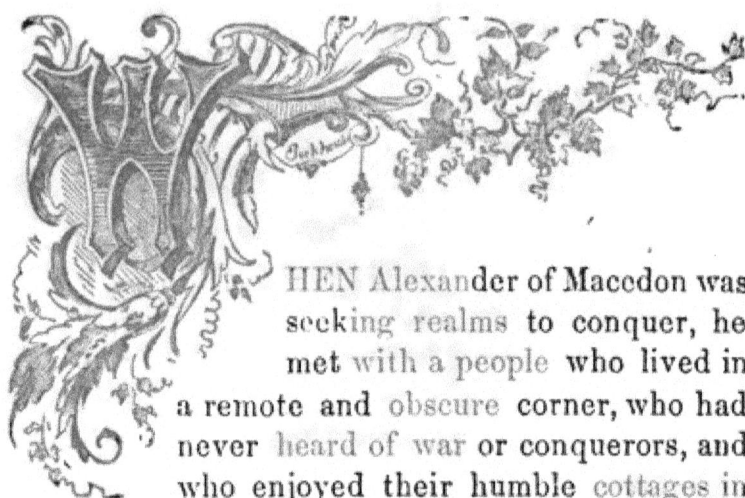

HEN Alexander of Macedon was seeking realms to conquer, he met with a people who lived in a remote and obscure corner, who had never heard of war or conquerors, and who enjoyed their humble cottages in profound peace. They met the Macedonian king, and conducted him to the dwelling of their ruler, who received him hospitably, and set before him, as a feast, dates, figs, and other fruits, made of gold.

"What! do you eat gold here?" asked Alexander.

"No ; but we imagined thou hadst food enough to eat in thine own country, and that it was a desire of gold that led thee forth from it. Why, therefore, hast thou come to us from so far a country?"

"It was not for your gold I came," replied Alexander ; "but I desired to learn your customs."

"Even so ; then abide among us as long as thou wilt."

While the ruler and the Grecian were conversing, two men of the tribe came in, to appeal to the ruler's judgment. The complainant spoke :

" I bought a piece of ground from this man, and when I was digging it found a treasure. The treasure is not mine, for I purchased only the ground. I never included in the purchase any hidden treasure, but this man who sold me the land refuses to receive the treasure from me."

The defendant now replied :

" I am as conscientious as my neighbor. I sold him the ground, and everything that might be in it ; therefore, the treasure is justly his, and I cannot take it."

" The ruler took time to understand the case clearly, and then asked one of the parties :

" Hast thou a son ?"

" I have."

He inquired of the other :

" Hast thou a daughter ?"

" Yea."

" So, then, the son shall marry the daughter, and the young couple shall have the treasure as a wedding portion.

Alexander betrayed some emotion.

" Is not my judgment just ?" inquired the ruler.

" Perfectly just," returned Alexander, " but it surprises me."

"How, then, would the case have been decided in thy country?"

. "To own the truth," said Alexander, "both the men would have been taken in custody, and the treasure seized for the king."

"For the king!" said the ruler, full of astonishment; "does the sun shine in that land?"

"Surely."

"Does the rain fall?"

"Of course."

"Wonderful! but are there gentle, grazing animals there?"

"There are, and of many kinds."

"Then," said the ruler, "it is for the sake of those innocent animals that the all-merciful Creator permits the sun to shine and the rain to fall upon your land; *ye* deserve it not."

"COME, sonny, get up," said an indulgent father to a hopeful son, the other morning— "Remember that the early bird catches the first worm!"

"What do I care for the worms?" replied the hopeful, "mother won't let me go a-fishing."

MONKEY LUCK.

CATS-PAW.

SEE the saucy rogue! How imprudently he laughs at the joke he is perpetrating on the poor helpless cat. The nuts are in the fire, all roasted, and ready to burn. Jocko wants them, and will have them, but don't mean to burn his own delicate fingers, by pulling them out. So he promises Miss Puss a liberal share of the delicacies, if she only lends him her paw to take them out of the fire. Puss demurs, and screams vociferously, but all to no purpose. She is in the scrape, having helped him to steal the nuts, and now she must bear the penalty of being in bad company. Puss is sadly burned, so that she cannot eat a morsel, and Jocko takes the entire spoil to himself, chuckling over his ready wit and good fortune. Look out sharp, boys, for the company you keep.

THE ROYAL TIGER.

THE Royal Tiger is a native of India, though sometimes found in the surrounding countries. It inhabits the low lands and jungles which are covered with briers, and thick shrubs, so compact as to be almost impassable to man. Its ferocious nature, extraordinary beauty, and great power, are its prominent characteristics, and give it a prominence among the whole animal creation. But though strong and powerful enough to defy man or beast, yet it prefers to get its prey by a stealthy attack, rather than by an open and bold assault.

Those who visited Batty's menagerie in Dublin, will remember that he had two lions and a tiger tamed together in the same cage, and whilst exhibiting at Roscre, a few days ago, the keeper of these animals, whilst in the cage with them, missed his foot and fell upon the tiger, which was asleep at the time. The animal became enraged, and jumping up, caught the unfortunte man by the thigh. A thrill of horror pervaded the hundreds of spectators who were visiting the exhibition at the time, and the man's destruction was deemed inevitable ; when to the inexpressible joy, as well as amazement of all present, the lion seized the tiger by the neck, and caused it at once to relinquish its hold, whilst the man was dragged out of his cage bleeding in a dreadful manner. He was immediately placed under the care of a doctor, and after a long and severe illness finally recovered.

A VISIT TO A MENAGERIE.

AT the entrance there was a band playing to attract attention, and a crowd of boys gathered round, looking at the stuffed ostrich at the window, and at the privileged persons admitted inside. As soon as we got in, there was the strong menagerie smell, which is unavoidable, however clean the animals are kept. On one side of the room was a row of cages ; over the first was printed "Royal Tigers," and royal looking animals were the two splendid beasts inside. They lay stretched out at full length, showing to advantage their beautifully striped skins. After admiring them for some minutes, we turned to the next cage, the African lion's. The

king of beasts was walking restlessly up and down
his little space, so different from the vast deserts
he had been accustomed to roam in his native
country. He was about the size of the tigers, I
thought, but shorter and thicker built. I think
even if one had not heard so much of the lion's noble, king-like appearance, he would at once single
him out as the most noble animal of all. I don't
think too much can be said of his beauty. His face
was so full of expression, and his great mane, mixed
with black, gave him a most majestic appearance.
A Java tiger was in the next cage ; and curled up
in the next, lay a creature looking like a great black
cat, but two or three times as large. This was a

THE LEOPARD.

black leopard; the man
poked at it with his stick
to make it get up, but it
only snarled, showed its
teeth, and caught at the
stick. The man said it was
a most fierce creature, and quite as dangerous as
the larger animals. We next came to two lionesses.
Their ladyships lay stretched out, one at each end
of their cage, half asleep. Certainly the animals
seemed rather lazy here, but perhaps they had
caught the city custom of keeping their beds late,
and hadn't yet aroused themselves, though it was
after ten. They differed from the lion in being
smaller, lighter built, and not having manes. The

Brazil tiger, or jaguar, looked more like a leopard than a tiger, but is larger ; he, too, was lying down.

THE HYÆNA.

The next comer, the hyæna, couldn't be accused of laziness ; he was walking round his cage with most praiseworthy industry, going as if the fate of the nation or his dinner depended on his getting round quick enough. As he comes tramping along, not heeding where he goes, he keeps running his nose against the wall, and the bars of the cage. His hide is a dirty yellow, or yellowish brown, with brown spots over it. He was certainly an ugly looking creature, and he didn't look the better for having the handsome jaguar on one side, and a lion on the other. I have always thought the hyæna a perfectly untameable animal, but reading an account of him when I got home, I found this passage : "It is a common, but erroneous idea, that the hyæna is wholly savage and untameable. Both species have been tamed, and instances are recorded of their manifesting all the attachment of a dog. The striped hyæna has recently been domesticated in the Cape territory, and is considered one of the best hunters after game, and as faithful as any of the common domestic dogs." We now came to the Bengal or Asiatic lion, who differs from the Cape or African lion, in being smaller, and of

a lighter, more uniform color. The man poked at
him with his stick, until, after some resistance, he
got up and began walking his cage. How dignified
and majestic he looked as he walked up and down !
It seemed a shame to poke at and torment the noble animal. And he looked nearly as well lying
down as standing up. Among the last animals in
the row were two bears. One of them was indulg-
ing in a strange exercise, viz :
moving his four feet in succession, first forward, or partly
sideways, and then back again, with the regularity of
a machine. This strange habit is peculiar to bears ;
they will do it for hours together. Bruin stood
with his back to the company, very impolitely, and
with his head down, seemed absorbed in watching
his toes. He was evidently practising the steps of
some cotillion, although he was the last individual
from whom dancing would be expected. Once he
turned round, looked gravely at the company a moment, as if expecting applause, and again became
absorbed with his steps. But now we heard sounds
from the lion's cage, and turned back to it. His
majesty, it seems, could not disguise his dissatisfac-
tion towards his visitors, and favored them with a
series of sounds, half growl, half yawn, as if to hint
to them that their presence was wearisome to him.
Such utter, intense disgust, the sound, and the very
look of his face expressed ! We were very much

amused. But leaving him, we went to the upper
part of the room, where were the cages of the
trained animals that Herr Driesbach goes in among
every day. There was also a blue-faced baboon.
This creature is found on the Gold Coast, and in
several other parts of Africa, and also in the East
Indies. He is the largest of the baboon kind, being from three to five feet high. He is a horrid,
ugly, disgusting looking creature, seeming to be
possessed of the concentrated ugliness of all the
monkey and baboon tribe. His cheeks are of a
deep blue color, and have no
hair on them. A narrow,
blood-red ridge extends down
the middle of his face, and terminates in the nose, giving him
an ugly enough appearance.
As he is entirely brutal and
untameable, and of enormous strength, he is of
course a dreadful creature to come in contact with.
But now again hearing sounds from the lion, we
repaired to his cage. As we passed the cages, we
saw Bruin still practising his dancing, and the per-
severing hyæna still tramping round, catching his
nose. The unhappy lion had evidently been re-
flecting over his grievances—his being stared at,
poked at, and poked up—both his person and his
temper—for the amusement of visitors ; and he
now gave full utterance to his indignation and

wrath in a series of roars. But one of the keepers
came up, and with his loud voice and stick soon
silenced him. Poor fellow! After enduring so
much, not even to be allowed the satisfaction of
grumbling at it, was the greatest wrong of all, and
the hardest to bear. But as there was an elephant,
rhinoceros, and llama down stairs we left the injured lion to see them. Every one has read about
the elephant, so I will not describe him, but merely
state a few facts I saw in the newspaper about this
particular elephant. He is
the largest in America, being
eleven feet high, and weighing
11,000 pounds. He is forty
years old; that, however, is
not old for an elephant; they
are said to live to one hundred
years, and some say to three or four hundred.
Every day he eats 400 pounds of hay, three bushels of oats, and drinks four barrels of water; then
he also gets a good many cakes, apples, etc., from
his visitors. There's a glutton for you! The rhinoceros is, perhaps, the most worth seeing of all the
collection. They are very hard to keep alive in
this country. One reason is, in their own country
they are accustomed to live by rivers, and in marshy
places, where they wallow a good deal in the mud;
and in this country, they must miss the water very
much, kept in cages as they are. This one looked

not unlike a great hog, particularly as he had no horn, it having been broken off. Their horns are often from three and a half, to four feet high. Think what a weight to carry on the nose. The largest rhinoceros' have, I read, nearly as great bodies as elephants, but having such short legs they do not appear so large. In the next pen there was a llama, a pretty, delicately formed animal, quite a contrast to the great, bulky rhinoceros. You all know how useful she is in South America, carrying burdens, so I will not enlarge upon that. This one was about four feet high, with long reddish hair on her body, slender legs, and a small pretty head. She seemed quite tame and gentle, letting me pat her head, and eating cake and apple from my hand. But when I had no more to give her, she was much displeased, stamped her foot, put up her head and spit in my face. A grateful return, certainly! Her ladyship might have found some more delicate way, I should think, of expressing her displeasure. The llama is generally gentle and docile, but if ill-treated, she becomes spiteful, and uses this strange mode of retaliation. We went down stairs to see Herr Driesbach enter the animals' cage. Before he went in, he passed through the menagerie looking at the animals. When he came to the cages he was to enter, the animals immediately began to jump about, as if they were delighted to see him. The lioness particularly, seemed transported

THE ZEBRA.

with joy ; she jumped up against the bars of the cage, threw herself on the ground and rolled, and testified in every way the greatest joy at seeing him. It evidently *was* joy, not fear. The partitions that divided the cages were soon taken away, and gas was lighted round the cage, which added much to the effect. A few minutes after, Herr Driesbach entered the cage. The performances were wonderful. He made the animals come to him, lie down beside him, jump up, on, and over him, open doors, and do several other things. It was wonderful to see these naturally fierce animals playing and leaping about him, obeying his every command, even letting him examine their teeth, and

CAMEL LEOPARD

put his head in the lion's mouth. I don't think there is any cause at such exhibitions for being nervous and frightened for fear of their getting provoked and hurting the man, for though we know them to be fierce, treacherous creatures, the man must know it still better, and know how to be on his guard against them. He would know how far he could go, without rousing them, and how far to insist on their obedience. Before I saw them, I thought there must be great cruelty used to subdue them so, and that they must be kept in most abject

fear, but this was evidently not the case ; they
seemed perfectly free and familiar, only having
such wholesome awe of their master as would make
them fear to disobey.   I do not see why there should
be greater cruelty used towards them than is used
to a dog ; who, though fond of his master, fears to

disobey him, knowing he will be punished ; and as these animals' training
begin when they are cubs, I should
not think it would be very hard to
keep them in proper restraint.   After
the performances we left, and I really
felt quite a tender feeling for the

THE OSTRICH. dear old hyæna, blue-faced baboon;
and all the other animals, and felt quite sad to think
I might never see them again.

HIPPOPOTAMUS HUNTING.

TO cut a supply of wood for a whaling cruise is a
work requiring some days, and often even weeks,
and it had been determined that the first, and if
need be the next day likewise, should be devoted
to a thorough inspection of the facilities of the

place, in order that we might work at as little dis-
advantage as possible.

Consequently we, the mate's boat's crew, had
been ordered to prepare for a general cruise. We
provided ourselves with a store of bread and beef,
filled the boat's breaker with water, spread our sail
to the breeze, and pointed the boat's bow toward
the nearest island. Landing here, we found nought
but a wilderness of low jungle, which was scarcely
penetrable, together with a poor landing. We examined three or four of the islets, and having at
last fixed upon a suitable place where to commence
operations, were about to return on board, when
the mate said:

"Trim aft, Tom, there's a good breeze, fair coming and going, and we'll take a look at the main-
land." Accordingly, the boat's head was laid shore-
ward, and we spread ourselves out at full length
upon the thwarts, enjoying an unusual treat of
some cigars which our chief officer had good natu-
redly brought with him.

When within about a mile and a half of the main-
land, we found the water shoaling, being then not
more than three fathoms—eighteen feet—deep.

"I saw black skin glisten in the sun just then,"
said the boat-steerer, who was aft, the mate having
stretched himself upon the bow-thwart to take a
nap.

"It was nothing but a puffing pig," said he,
drowsily.

" There it is again, and no puffing pig either—
nor—no," said he, with some degree of animation
—" nor any thing else that wears black skin that I
ever saw before."

This had the effect of rousing us up, every one
casting his eyes ahead to catch a sight of the ques-
tionable " black skin."

" There he blows !"—" and there again !"—" and
over here, too," said several voices in succession.

" It ain't a spout at all, boys ; let's pull up and
see what it is !"

We took out our oars, and the boat was soon
darting forward at good speed toward the place
where we had last seen the object of our curiosity.

" Stern all !" suddenly shouted the mate, as the
boat brought up " all standing" against some object which we had not been able to see on account
of the murkiness of the water, the collision nearly
throwing us upon our backs into the bottom of the
boat.   As we backed off, an enormous beast slowly
raised his head above the water, gave a loud snort,
and incontinently dove down again, almost before
we could get a fair look at it.

" What is it ?" was now the question—which no
one could answer.

" Whatever it is." said the mate, whose whaling
blood was up, " if it comes within reach of my iron,
I'll make fast to it, lads—so pull ahead."   We were
again under headway, keeping a bright look-out for
the reappearance of the stranger.

" There they are, a whole school," said the mate, eagerly, pointing in shore, where the glistening of white water showed that a number of the nondescripts were evidently enjoying themselves. "Now, boys, pull hard, and we'll soon try their mettle."

" There's something broke water, just ahead," said the boat-steerer.

" Pull easy, lads—I see him—there—way enough —there's his back !"

" Stern all !" shouted he, as he darted his iron into a back as broad as a small sperm-whale's.

"Stern all—back water—back water, every man!" and the infuriated beast made desperate lunges in every direction, making the white water fly almost equal to a whale.

We could now see the whole shape of the creature as, in his agony and surprise, he raised himself high above the surface. We all recognized at once the Hippopotamus, as he is represented in books of natural history.

Our subject soon got a little cooler, and giving a savage roar, bent his head round until he grasped the shank of the iron between his teeth. With one jerk he drew it out of his bleeding quarter, and shaking it savagely, dove down to the bottom. The water was here but about two fathoms deep, and we could see the direction in which he was traveling along the bottom, by a line of blood, as well as by the air-bubbles which rose to the surface as he breathed.

SPEARING HIPPOPOTAMI.

"Give me another iron, Charley, and we'll not give him a chance to pull it out next time."

The iron was handed up, and we slowly sailed in the direction which our prize was following along the bottom.

"Here's two or three of them astern of us," said the boat-steerer.

Just then two more rose, one on either side ot the boat, and in rather unpleasant proximity ; and before we had begun to realize our situation, the wounded beast, unable any longer to stay beneath the surface, came up to breathe just ahead.

"Pull ahead a little ; let's get out of this snarl. Lay the boat round—so—now, stern all !" and the iron was planted deep in the neck of our victim. With a roar louder than a dozen of the wild bulls of Madagascar, the now maddened beast made for the boat.

"Back water !—back, I say ! Take down this boat-sail, and stern all ! Stern, for your lives, men !" as two more appeared by the bows, evidently prepared to assist their comrade. He was making the water fly in all directions, and having failed to reach the boat, was now vainly essaying to grasp the iron, which the mate had purposely put into his short neck, so close to his head that he could not get it in his mouth.

"Stick out line till we get clear of the school, and then we'll pull up on the other side of this fellow. and soon settle him with a lance."

This was done ; and as we again hauled upon the still furious beast, the mate poised his bright lance for a moment, then sent it deep into his heart. With a tremendous roar, and a desperate final struggle of scarcely a minute's duration, our prize gave up the ghost, and after sinking for a moment, rose again to the surface, lying upon his side, just as does the whale when dead.

His companions had left us, and we now, giving three cheers for our victory, towed the carcass to the not far distant shore. When we here viewed the giant, and thought of the singular agility he had displayed in the water, we could not help acknowledging to one another that to get among a school of Hippopotami would be rather a desperate game.

A SPIDER DROWNED OUT.

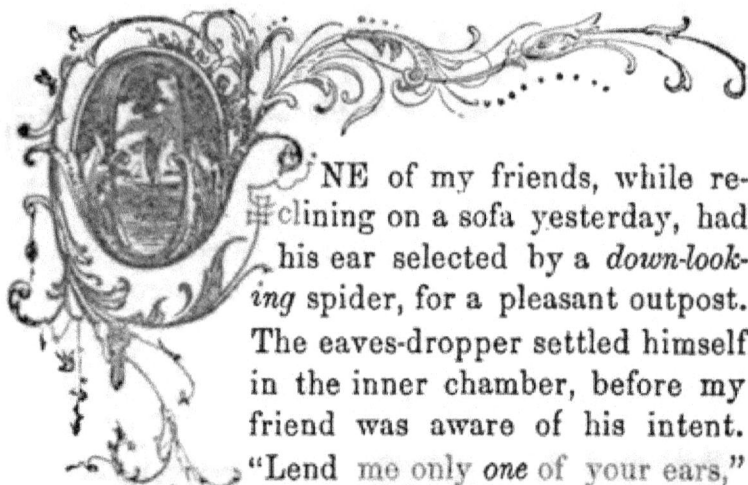

NE of my friends, while re-clining on a sofa yesterday, had his ear selected by a *down-look-ing* spider, for a pleasant outpost. The eaves-dropper settled himself in the inner chamber, before my friend was aware of his intent. "Lend me only *one* of your ears," said the considerate intruder. The question arises, what was his object? Did he mean to lure flies in-to his retreat, or to watch for and seize upon them from his hiding-place?

Possession is ten-tenths of spider-law; and he forthwith arranged (arraignée) his limbs, and fold-ed himself nicely up in his new lodgings, not partic-ularly to the comfort of the rightful owner thereof.

After several unsuccessful attempts to dislodge him, the bright thought of pouring in water was resorted to; and having a whole lake of that valu-able remedy, we spared not, and so floated Mr. Spi-der from the premises.

He speedily ran off to higher ground for his encampments, the flood being too much for him, and we—generously *let him go.* My friend, Uncle Toby-like, said, the world was wide enough, only, noses and ears must be left unmolested.

ICHNEUMON FLY.

Did you ever, my dear little observers, did you ever notice the plans and ways of this wasp-shaped insect? I never studied about him, and perhaps you can tell me more than I can tell you. I will tell you what I *saw.* I saw these little architects construct many houses of mud just large enough to hold one of themselves. No mason could build smoother, or construct an edifice piece by piece, so that you could not see the joining. They choose an upright board, or a roof of wood, sheltered from rain, against which to secure these homes for their children. Instead of a house, suppose I call it a cradle, for it bears that shape. Within each, at the extreme end, in a smooth little hollow, is placed the infant fly, in the form of an egg. Only one in every separate cradle. I never yet saw twins. Now, the

ICHNEUMON FLY. parents go spider - hunting. Having captured a good fat one, they put him asleep by magnetism, all their own, and place him carefully by the little egg. Then another and another, until ten are packed in.   Ten living spiders, but all lying dormant, fill the nest.

Then **the open** door **at** the top is shut **tight and sealed with the** same **mud** material of **which the cradle is** composed.

After many days, a nice young grub awakes **to** eat, and there is his food all prepared. He begins at one spider, and **by** the time he has eaten the ten he **is so** stout and well-grown, he commences to **break** out of jail, as **it** were. And he *does* come forth.

When his fly life begins, what a joy it must be ! **How he spreads** his gauzy wings in the **sun, and** hums his delight. **He has all the happiness he is** capable of receiving. Watch him, admire his beautiful organization. See that slight thread of connection, through which all the nerves and all the digestive powers exist.

"God is **good," is** the **written word to be read in** the life **of every living** thing.

DEACON SHORT'S CATTLE.

ACCORDING to the best of my knowledge and belief, all horned cattle, so far as their habits and manners are concerned, are very much alike. Deacon Short's cattle, I am persuaded, would not behave at all differently from Squire Long's cattle, in similar circumstances. It may be worth while, nevertheless, to notice how the deacon's cattle acted on a certain occasion.

I must first say, however, that Deacon Short was a merciful man, and, therefore, was merciful to his beasts. No living thing around him ever suffered for the want of care. If he thought there was a lack of comfort anywhere, he could not feel comfortable himself. Accordingly, at the time referred to—that is, near sundown, on a bitter cold day in January—he might have been seen about the

shed where his cattle were quartered, making liberal additions to their straw beds. "Of course, then
they had a good comfortable night's lodging."
Don't be too sure of that, my friend. They might
have slept much more comfortable than they did,
had they been a little more accommodating. There
was old Brindle, in particular. She pushed her in-
feriors about without mercy, and seemed to care
a great deal more for herself than for all the rest
together. She meant to have the very best place
under the shed, if she could find where it was.
Young Spot, too, gave signs of a determination to
do as well as she could. The good deacon was
quite displeased to see them treat their companions
with so much rudeness, and, to teach them better
manners, gave them two or three pokes apiece with
the tail of his pitchfork. All that amounted to but
little, however. As soon as he was fairly out of
sight, they began to perform as before. Any steer
or heifer that ventured too near them got a cruel
thrust of their horns for being so imprudent. They
finally took possession of that part of the shed
which seemed, on the whole, most desirable. They
could not agree, however, to lie down very near
each other. The rest had more kindly feelings
among themselves, and huddled down together in a
remote part of the shed.

In the morning when Deacon Short entered his
barn-yard, there stood Spot and Brindle shivering

with cold. The wind had changed during the night and whirled in upon them a pretty thick covering of snow ; while the rest of the herd, by lying close together, had, in a measure, protected and warmed one another. The deacon understood the case at once. "You selfish old creature," said he, addressing himself to Brindle in particular, as if she understood English just as well as she did her own cow-language. "Good enough for you. You would have lost nothing, you see, being a little more accommodating. Had you been willing to warm your neighbors, they would have warmed you as much in return." Then turning his reflections into a talk to himself, he went on : "So it is. While selfishness is sure to get punished in some way, kindness and benevolence are as certain to meet with reward. A man is accommodating himself even while he is accommodating his neighbor. He who does good to another, does good to himself at the same time."

COMMUNICATION OF IDEAS AMONG CATTLE.

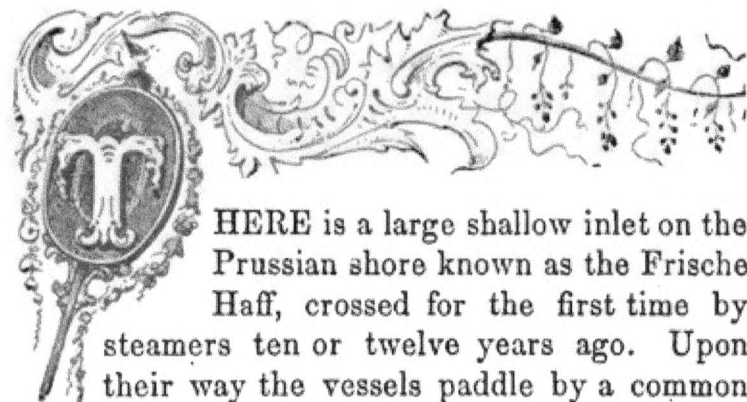

HERE is a large shallow inlet on the Prussian shore known as the Frische Haff, crossed for the first time by steamers ten or twelve years ago. Upon their way the vessels paddle by a common near the Elbing river, upon which the towns-people turn cattle out to graze. When the first steamers passed this common, they caused every flank of beef to quake; such fiends in dragon shape had never appeared before to try the nerves of any cow, or to excite wrath in the bully bosom of the most experienced among the warriors of the herd. With tails erect, therefore, and heads bent down, the whole colony upon the common charged over dykes and ditches inland, roaring horribly. Every appearance of the steamer, to the great joy of the crew, caused a panic and a scattering of oxen, until, after a few days, the animals had become hardened to the sight, and took it as a thing of course, which meant no harm to them. Now, all the horned beasts on the common during that first year were in the usual way to be fatted. In the following spring they had gone the way of beef, and their place was filled by a new generation altogether. So soon, therefore, as the Haff was clear of ice, and the steamers began to ply daily upon the route between Elbing and Konigsberg, the sailors were on the alert again to witness the old scene of uproar by the water side. But they were disappointed. Though there were the pasture ground well stocked with new recruits for the market, who had come from distant inland farms or out of stalls within the town, though scarcely one of them—if any one—had ever seen the apparition of a steamboat, not a cow flinched. The members of the whole herd went on grazing or stared imperturbably at the phenomenon. It was a new thing, no doubt, for them to see. Every spring the first passing of the steamers is in this way regarded by a fresh generation on the common with complete indifference. The experience acquired by its forefathers ten or twelve years ago seems to be now added to the knowledge of every calf born in any corner of our province. And yet, in what way, have these calves been educated ? or, if this fact has been taught to them at all, what else may they not know !

DR. DUNLOP AND THE TIGER.

THE Dr. while in the East Indies conquered a royal tiger with a bladder of Scotch snuff. Having crossed the river Ganges with his quarterly allowance (seven pounds) of snuff, he observed a tiger at some distance. Being without guns, he ordered his men to use their oars as weapons of defence. They formed into a close column, with our backs windward, while the doctor emptied the contents of the bladder into a piece of canvass, and danced upon it till it became as fine as dust. The tiger continued winding, and occasionally crouched. When he approached within twenty yards of the party, the doctor discharged about half a pound of the ammunition, part of which was carried by the strong wind into the face of the tiger, who growled, shook his head and retreated. In a few minutes he returned to the charge, approaching the party cautiously, and rubbing his eyes occasionally with his fore-legs. When within about fifteen yards of the party he again crouched, and as he was preparing to make his murderous spring, the doctor and his party let fly at him about two pounds of snuff, which told well, for the royal tiger commenced roaring, and springing into the Ganges, fled to the opposite shore.

DUELLING AMONG MOSQUITOES.

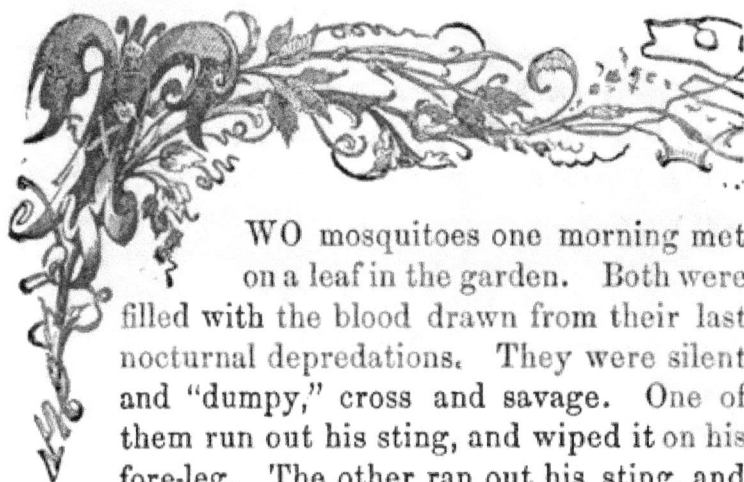

WO mosquitoes one morning met on a leaf in the garden. Both were filled with the blood drawn from their last nocturnal depredations. They were silent and "dumpy," cross and savage. One of them run out his sting, and wiped it on his fore-leg. The other ran out his sting, and pointed it towards the first musquito. This was considered an insult. And so the offended mosquito steps up to the other, and says :

"Did you turn up your sting at me ?"

The answer was—"I run out my sting ; you can apply it as you choose."

"Sir," says the first, "you are very impertinent."

Answer—"Sir, your remark savors of rascality."

"Ha," exclaimed the other, "a downright insult ! No gentleman mosquito will submit to such treatment without demanding satisfaction ! Draw, villain, and defend yourself at once !" They rushed together, and running one another through the body died "honorable" deaths.

If anybody is disposed to question the honor of these two mosquitoes, or from their conduct to impute any dishonor on their race, it should be said, that they were not bona fide, uncontaminated, and unsophisticated swamp mosquitoes, but that they had been lurking about a boarding-house, where they had learned something of *polite* society, and had acquired some *uppish* notions that made them feel very grand.

THE RABBIT.

A FABLE.

A rabbit young—more weak than keen,
Held in its mouth a walnut green;
His parents told it, "nuts have meat—
The kernel of that nut is sweet!"
But nibbling its green coat uncouth,
The ignorant creature doubts the truth;
Deems it unfraught with meat or bread,
And gives no heed to what is said:
His wisdom teeth were still uncut;
The youngster threw away the nut.
A keen-eyed monkey watched the lad—
Seized the same nut extremely glad;
Held it compress'd with dextrous paw,
Then fairly cracked it with his jaw;
Rejoiced and grinning o'er the treat,
Breakfast upon delightful meat.
Then young Rabbit says with sneer:
"Your parents told the truth my dear;
But idle boys, with giddy stem,
Knowledge is never to them."

THE INDIAN LIZARD.

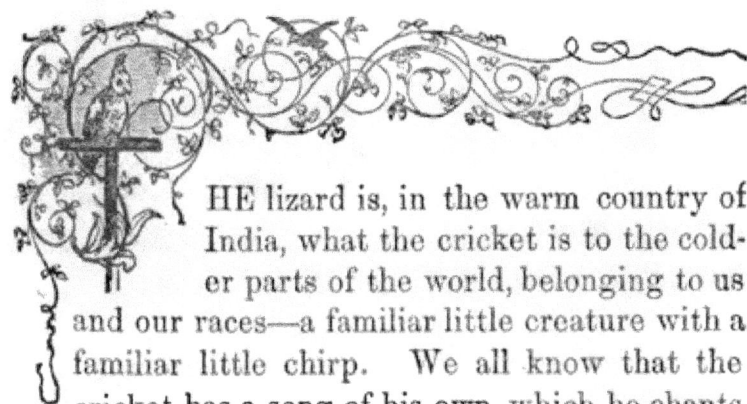

HE lizard is, in the warm country of India, what the cricket is to the colder parts of the world, belonging to us and our races—a familiar little creature with a familiar little chirp. We all know that the cricket has a song of his own. which he chants when the hearth is cosy. Many have pleased themselves in listening to it, and sometimes making out meanings for it. Mr. Dickens once heard a cricket singing against a tea-kettle. The kettle began it, as everybody knows.

These crickets and lizards are, in fact, members of a very large family to which fanciful people have at all times been extremely partial. The little grass-hopping folks are spoken of by those who have written earliest in the world, that is, the Hebrew prophets and singers; the Greeks had an idea they were born from the soil. For which reason the beautiful maidens of Greece, who could boast their descent from a long line of ancestry, in their own country, used to wear golden grasshoppers, or cicadas, in their hair, as much as to say—

"we have the best and noblest pedigrees on this ground." Greek poets have made cheerful and loving odes to the cicada—one of the musical brotherhood, in fact, only of a kind of lower order—a songster that always reminded them of the fine weather and soft breezes, and the summer sports and enjoyments under the shade of trees. Bards of other countries, too, have made merry or tender allusions to it. Lamartine has a melancholy little ode to his cricket, and Lord Byron speaks of

> The shrill cicadas, people of the pine,
> Making their summer lives, one ceaseless song.

Those crickets and grasshoppers are as well known to us, as to any other people, and we find they are indeed almost incessant singers in the genial season. Sometimes, at night, when all other sounds are still, they fill the air with their chirpings, being then, doubtless, performing their oratorios, concerts, operas and *charivaris*, all together in the open air.

But I began with the lizard and must not forget it. As I was saying, the lizards are household creatures in India, loving the open windows and verandahs, as their Western cousins love the warm inglenook. Many stories are told about them. The natives say a benevolent lizard will watch the house at night, and make a rousing noise, if robbers try to break in. But a wickedly disposed lizard will

actually encourage the villains, and come forward
to show them where the money is locked up. Some
of the natives say they understand the talk of the
izards as they see them in groups of parents and
children, on the verandahs. That is a very old
fancy of the Eastern people—the power of understanding the speech of the speechless creatures.
A little boy one day, in a bungalow, near Madras,
told some European officers that he heard one lizard say to another, outside the window, "My wife
is coming this evening!" they laughed at the lad
and one of them cuffed him for telling lies. After
dinner, a ramper of wine came from Madras, and
when it was opened, out jumped a lizard, and the
same little boy heard the other shouting away on
the verandah : "Here she comes, tak, tak, tak! I
knew she'd be here, tak-a-tak-a-tak!" The unbe-
lievers then begged the little fellow's pardon and
gave him some sweetmeats to comfort him. Such
is one of the lizard stories told and believed by the
natives of that part of India.

THE WHALE.

HALLO! old fellow, laid high and dry,
    Upon a cake of ice;
Methinks you have found that "getting high,"
    Is not a convenient vice;
And that "half-seas-over," as there you lie,
    Is any thing but nice.

You'll doubtless protest, though the doctors still
    The contrary declare,
That being kept dry against one's will,
    The health is sure to impair ;
And it's quite as bad as an arsenic pill,
    For a whale to "take the air."

But, where are you bound, in your flat-bottomed smack,
    Without rudder, mast, or sail ?

Do you take old England in your track,
    And call on the Prince of Whales?
Will you stop at New York, as you go back,
    And with Governor Fish regale?

You need not fear your craft to steer
    Over Nantucket shoal;
Nor deem it unsafe approaching near
    New Bedford or Holmes' Hole;
Nor that Judd or Macy, or any one here,
    Will tap your brains for toll

Whale oil is no longer in vogue, you know
    We're quite in another line.
Camphene, kerosene, *et cetera*, now,
    Have taken from you the shine;
We get our light from shote and sow,
    And the *sperm o' the city* is—swine.

But, hark'ee, old fellow, don't flap your jibs
    In **Paris, or Broadway, please**;
There's a terrible rage among our "ribs"
    For skirts of ample degrees;
And the ladies will tear you all to flibs,
    Your *bony parts* to seize.

RATS.

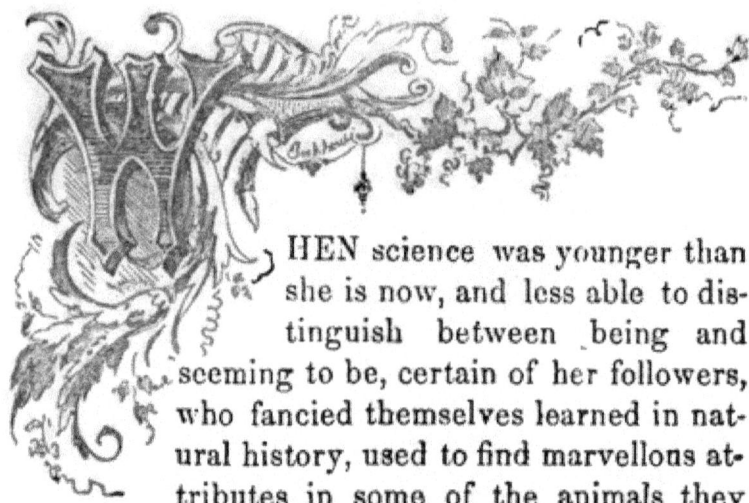

HEN science was younger than she is now, and less able to distinguish between being and seeming to be, certain of her followers, who fancied themselves learned in natural history, used to find marvellous attributes in some of the animals they wrote about. For reasons not easy to discover, they seldom mentioned rats without expressions of fear or abhorrence, giving the creatures credit for more than human intelligence. There was no wickedness that rats were not willing to perpetrate. Then there appeared to be strange relations between the cunning rodents and human beings, investing them with a mysterious character, not only in the eyes of the multitude, but in the opinion of students. At times they were more than half suspected to be agents of the Evil One.

Southey, in his *Doctor*, remarks that whatever man does, rat always takes a share in the proceedings. Whether it be building a ship, or erecting

a church, digging a grave, ploughing a field, storing a pantry, taking a journey, or planting a distant colony, rat is sure to have something to do
in the matter ; man and his gear can no more get
transported from place to place without him, than
without the ghost in the wagon that "flitted too."
How is it that a rat knows when a house is about
to fall, or a ship to sink? Where did they learn
to carry eggs down stairs, from the top of the house
to the bottom, without breaking! Who taught
them to abstract the oil from long-necked flasks, by
dipping their tails in, and then licking the unctuous
drops from the extremity? What precedent had
they for leading a blind companion about by a
straw held in the mouth, and how did they know
he could not see? All those are questions requiring no small amount of ingenuity to answer.

As with nations, so with rats ; one tribe comes
and dispossesses another. The rats that used to
gnaw the bacon in Saxon larders in Alfred's reign ;
that squealed behind the wainscot when Cromwell's
Ironsides were carrying royalist mansions ; that
disturbed the sleep of George I., were a hardy
black species, now seldom seen, and doomed, apparently, to become as rare as the dodo. Like the
Red Men in presence of the Palefaces, they have
had to retire before the Norwegian rat, larger in
size, and brown in color. Notwithstanding all the

popular notions on the subject, it is difficult to explain why this was called the Norwegian rat; for it did not come from Norway. It may surprise those, who are sticklers for their Scandinavian origin, to know that this rat was brought to England from India and Persia, in 1730.

In 1750, the breed made its way to France; and its progress over Europe has since been more or less rapid. When Pallas was traveling in Southern Russia, he saw the first detachment arrive near the mouth of the Volga, in 1766. The species multiplies so rapidly, breeding three times a year, each litter numbering from twelve to twenty, that a single family, if kept from harm's way, would produce nearly a million in two years. No wonder they drove out our aboriginal black rat! In Ireland, they did more; they killed the frogs, once numerous in that country; and since the diminution of the croaking race, the waters, as peasantry say, have been less pure than formerly. The Isle of France was once abandoned by the Dutch, because of the prodigious increase of rats; human life was hardly safe from their attacks.

After making themselves comfortably at home in England, the country of their adoption, they sent colonies across the Atlantic—rat empire, like men's empire, taking its course westward. In the West Indies they found congenial quarters, no cold, and plenty of food; and, multiplying in consequence at an astonishing rate, they became a destructive and intolerable pest, until the inhabitants were obliged, in self-defence, to poison them with arsenic and pellets of cassava. The remedy was attended by dismal results, for tormented by thirst, after eating the poison, the rats swarmed down to drink at the streams, and falling in, the water was poisoned, and a great mortality followed among the cattle that drank from the same rivers.

Besides this check, they have many natural enemies in the islands; the *Fermica imnivora* is not the least formidable; a battalion of this species, known as the Raffle's ant, makes but short work of clearing a plantation of every rat. At one time the negroes used to catch the rats, and expose them for sale in the markets of Jamaica, where the black population were always willing purchasers. The Chinese, too, have a weakness for "such small deer;" and it is a standing bit of fun on board ships lying in Canton harbor, to catch a rat, and hold the struggling animal up by the tail, in sight of the celestial crews in the tea-lighters alongside. A shout is immediately set up, and no sooner is the rat flung from the ship than an uproarious scramble follows for the possession of the coveted prize.

The Greeks knew a good many things; but if naturalists are to be believed, they did not know either the Norwegian rat, or the black rat : a large sized mouse was their familiar pest. Where the

black rat originally came from, is a mystery. Some
suppose it to be a native of America. But how did
it get to Europe? Did it cross the Behring's Strait
and traverse the whole continent of Asia? One
cause of its present rarity, besides the invasion
mentioned above, is, that it brings forth not more
than five or six young at a time, and only once a
year.

There are about one hundred species of rats,
large and small, audacious and harmless ; very few,
however, devoid of the mischievous propensity.

Nine inches is a respectable length for a Norway
rat ; but the *giant rat* of Malabar is twenty-four
inches long—one half body, the other half tail.
The *hamster* species swarm in the southern pro-
vinces of Russia, and has settlements in Hungary
and Germany. They are excessively fond of liquor-
ice, whether wild or cultivated, and find abundance
of either in those countries, committing sad havoc
in the plantations.

For winter use, they store up in their burrows
from twelve to one hundred pounds of grain in the
ear, and seeds in pods, all well cleaned and dried.
The hamster is about the size of the Norway rat,
but with a tail not more than three inches in length.
It has a pouch in each cheek, not seen when empty,
but when full they resemble blown bladders coated
with fur. These pouches are the animal's panniers,
and are generally carried home well filled from fo-

raging expeditions, when they are emptied by
pressing the forepaws against them.   Dr. Russell,
who dissected one of these rats, found the pouches
filled with young French-beans, packed one upon
another, so closely and skillfully that the most expert fingers could not have economized the recep-
tacle to greater advantage.   When taken out and
laid loosely, they formed a heap three times the
bulk of the creature's body !   The hamster, moreover, is brave as well as prudent, and shrinks from
no enemy, be it man, horse, or dog ; mere size has no
terrors for it.   If facing a dog, the rat empties his
pouches of their contents, and then inflating them
to the utmost, gives such a big, swollen appearance to his head and neck, as to present a most extraordinary contrast to his body.

The two sexes live apart in their habitations—
the males in one set of chambers, the females in
the other ; a practice which again shows analogy
between rats and some human sects.   The peasants
dig down to the burrows in winter, and seizing the
stores of grain, and the torpid rats, they eat the
flesh of the latter in some places, and sell their
skins.   In Germany, rewards are given by the au-
thorities for all the rat skins brought in ; and it is
on record in the town hall of Gotha, that not fewer
than 145,000 were paid for during three seasons.

Somewhat similar in habit is the *economic rat*,
which is found inhabiting the American and Asiatic

shores of the Arctic Ocean. This species generally form their abode in a turfy soil, where they ex-
cavate chambers a foot in diameter, with a flat
arched roof, and at times thirty entrance-passages
ramifying in different directions. Besides the lodg-
ing-vaults, they dig others, to be used as store-
houses, and employ themselves during the summer
in filling these with edible roots ; and so careful
are they over the task, that if the least trace of
damp appears, they bring out the roots again and
again on sunshiny days 'till they are sufficiently
dried.

Like their German congeners, they are exposed
to pillage, especially in Kamtschatka, where the natives in winter often run short of provisions. They
are found also in Iceland ; but food being scant in
that inhospitable country, the *economic* foragers
have frequently to cross and recross rivers and
lakes in their search for provant. Olaffsen relates
that on such occasions "the party consisting of from
six to ten, select a flat piece of dried cow-dung, on
which they place the berries they have collected,
in a heap in the middle ; and then, by their united
force, drawing it to the water's edge, launch it, and
embark, placing themselves round the heap, with
their heads joined over it, and their backs to the
water, their tails pendant in the stream, and serv-
ing the purpose of rudders."

THE PET CHICKEN

HENRY'S father was a farmer, and had a great many hens and chickens. One morning, when Henry went out to assist in feeding them, he saw one of the little chickens whose foot had been injured, so that she was quite lame, and she could not run after the rest of the brood. Chickens do not show much affection for each other, and never seem to care much if one of their companions is hurt ; they probably do not know any better ; so they all ran off to some newly ploughed ground where there were plenty of worms, and left poor little Lamefoot to peep and hobble along by herself.

Henry took the little thing up carefully. Lamefoot peeped and screamed very loud, when she found herself held fast in Henry's hand, and struggled to get away, but she found that by struggling she only hurt her lame foot more, and so she concluded to lie still and bear confinement as patiently as she could.

Henry carried the chicken in and showed it to his mother. She put a little cold cream on the chicken's foot, and told Henry she thought if he could keep her from running about for a few days, she would get as well as ever. So Henry made her a little coop in a shady corner at the back of the house, and shut her up in it.

He took care to feed his little patient two or three times a day, and keep her well supplied with caterpillars, so that Lamefoot became quite contented with her situation. In a short time her foot became as well as ever ; but she had become so attached to her quiet little corner, that after she was able to run about everywhere, she always came back every night to roost in the little coop which Henry had made her.

And he became so fond of his little pet, that he used often to carry her out corn, or grain, or fruit, whatever he thought she would like, and she would come to him and eat out of her hand.

By and by Lamefoot grew up to be a great hen, and furnished Henry with a good supply of eggs, which he always ate with a better relish than any others ; and the next spring she brought him out a fine brood of chickens, of which she took such excellent care that they were considered the finest in the farm yard, and his mother was very glad to accept from Henry a couple of pair for her Thanksgiving Chicken Pie, when that joyful occasion came round.

THE PANGOLIN.

THE PANGOLIN.

WHAT do think of that, boys? Is that a fish, a beast, or a bird?

"I am sure I don't know. I should think it was some sort of a dog, with his forelegs cut off," says one.

"And I should think," says another, "it was a young crocodile, or something belonging to that family."

"And I," said a third, "I don't know what it is. I wish you would tell us."

Well, it is the *Pangolin*, sometimes known by the name of the scaly ant-eater, and a scaly looking rascal he certainly is. He is a native of Asia and Africa, and lives on ants. He has no teeth, but is armed, instead, with a long, thin like tongue, which he pushes into the narrow passages of the ant hills and draws out his victims with great ease. He does not seize them, or impale them, but his tongue being furnished with a thick, gummy saliva, the insects stick to it, and are drawn out easily.

This queer fellow seems to have but two legs, and so indeed he has ; but, a substitute for forelegs, which he does not need, he has, as you see, just under his head, a fierce array of nails, or claws, as if his legs were drawn in, out of sight. With these claws, which are strong and sharp, he can tear open the ants' nests, climb trees, and defend himself from his enemies.

The Pangolin has a very queer way of rolling himself up in a heap, with his scales all on the outside, so that even the hyæna and the tiger cannot hurt him. Sometimes, when he has climbed a tree in search of food, he saves himself the trouble of creeping down, by rolling himself into a ball, and dropping to the ground. The tail, with its pointed scales, is used to assist him in climbing. Sometimes, when going up a tree or a post, he will hold on by his feet and tail, and throw his body back, as represented in this cut, and swing himself to and fro, as if he enjoyed the exercise.

GAZELLES AND GAZELLE-HUNTERS.

THE gazelle is one of the most beautiful animals imaginable. Did you ever see one? Probably not. The gazelle is not a native of our country, and is very seldom brought here. I saw two or three at the famous Zoological Gardens in London, and I assure you they furnished me a great deal of amusement.

Of all the animals in the world, unless the poets deceive us, the gazelle has the most beautiful eye. You recollect what Thomas Moore says on that point, in one of the sweetest lyrics in the English language :

"Oh, ever thus from childhood's hour,
   I've seen my fondest hopes **decay**;
I never loved a tree or flower,
   But 'twas the first to fade away ;
I never nursed **a** dear gazelle,
   To glad me with its soft blue eye,
But when it came to know me well,
   And love me, it was sure to die."

Passing over the poet's unhappy mood **of mind,**
occasioned, probably, **as my good** old uncle Barna-
bas used to say, **by** eating rather too freely of un-
ripe fruit, from the little I have seen of the gazelle,
I don't know **that** its eye is overpraised **in this**
stanza. Still I think I have seen human eyes quite
as attractive. They were **to** me **at** all events.

The gazelle is a native of Asia and Africa. The
chamois with which I became quite familiar while
traveling in Switzerland, though it greatly resembles the gazelle, is not placed in the gazelle family.
There are some twelve distinct species of this animal, each differing but very little from the **rest.**
They have all small limbs in proportion to **the oth-**
er parts of **the body, and are well** adapted for running gracefully **and swiftly.** They have a cloven
foot, like **the sheep.** Their hair is short, but **fine**
and glossy.

In some countries where the gazelle abounds,
falcons are bred to capture them. The mode in
which the capture is effected is cruel in the extreme.
Whenever the hunters see a gazelle at the proper

distance, they let the bird loose. The falcon, with
the swiftness of an arrow, flies to the poor gazelle,
which is unable to escape. The talons of the bird
are fixed, one in the gazelle's cheek, the other in
its throat ; and the innocent creature is so faint,
from the loss of blood, that its pursuers overtake
it and kill it.

Sometimes, too, the gazelle is hunted by means
of the ounce, a very savage animal, which, however,
can be tamed so as to be perfectly docile. The
ounce sits on the horse with the hunter, and remains there, with the composure of a cat in the
chimney corner, until a gazelle is pointed out. Then
the fierce animal creeps along carefully, without
making any noise, until he comes within a few feet
of his prey, when he pounces upon him, and de- ~
stroys him almost instantly.

There is another way in which the gazelle is
caught. A tame gazelle is bred for the purpose,
which is taught to join a herd of wild ones, whenever it perceives them. The hunter places a noose
around the horns of the tame gazelle in such a manner that, when he comes in contact with the others
(for they invariably fight at such a meeting,) the
horns of the wild gazelle will be entangled in the
noose on the head of the tame one, in which case
the two fighters can't separate themselves.

Another mode of catching the gazelle is by means

of the *lasso*. The natives surprise the gazelles in a thicket, and then dexterously throw the *lasso* so that it is wound around the legs of the animal.

You see that in all these different ways of capturing the gazelle there is nothing that looks very like honorable warfare. If people should adopt the same methods of hunting the gazelle that are resorted to by the chamois-hunter, the gazelle might laugh at all the military tactics of his enemies. Give the gazelle a fair field, and he would most certainly win the day. His legs would be his salvation.

THE ELEPHANT..

THE elephant is the most sagacious and intelligent of all quadrupeds, and the nearest in its approach to human reason. Its enormous size and immense strength render it a formidable enemy when provoked, but even in a wild state it is not ferocious. It is easily tamed by kindness and caresses, and when properly treated it is obedient, grateful, and discriminating to a degree that proves it to be endued with a portion of something very similar to rationality. Elephants, even when wild, evince signs of great ingenuity, forethought and

memory ; and show much regard and consideration for each other. They generally go in herds or
companies ; sometimes carrying in their trunks
branches of trees which they use as fans to cool
themselves. If one of them gets hurt, the others
take care of him, bringing him food and nursing
him till he recovers. In crossing a river the old
ones swim over first, to seek a proper landing place :
and when safe on the other side, give a signal, by
a sort of cry or shout, for the young ones to follow.
The little elephants then venture across, supporting each other by interlacing or locking their trunks
together. The old elephants sometimes carry the
very small ones laid high across their tusks, twin-
ing their trunks round them to prevent their falling. If they find a dead elephant in the woods,
they stop and cover him with grass or with the
boughs of trees.

The elephant will eat almost every sort of vegetable food, and is extravagantly fond of confection-
ery, but abhors flesh and fish. I have seen them
drink wine and porter, taking the bottle in their
trunk, which they bend under to the mouth, holding back the head so as to let the liquor run down
their throat. In India the tame elephant is used
for various services. He will perform more work,
and carry or draw greater burdens, than six horses :
but he must be well fed and properly taken care
of. It is said that he will eat a hundred pounds of

rice in a day, drinking forty gallons of water : but his diet should be varied with fruit and herbage, and he must be led to the river twice a day for the purpose of bathing.

There is a story of an elephant becoming so fond of his keeper's child that he could scarcely bear to have it taken out of his sight. At last he would not eat his food unless the infant's cradle was placed between his feet, and as soon as this was done he ate heartily. If the child awoke and cried, the elephant frequently put it to sleep again by rocking the cradle with his trunk.

The Duke of Devonshire had a very fine elephant which he kept in the grounds of his villa at Chiswick, near London, in a handsome stone building of one story, erected purposely for the accommodation of the animal, who went in and out on a slanting platform or inclined plane. Some relatives of mine saw it there a few years since. The elephant was walking about under the trees. He seemed very proud of a rich mantle or pall of blue and crimson which was thrown over him. At the desire of his keeper he took it off with his trunk, spread it evenly on the grass, carefully smoothing every wrinkle, then folded it square and neatly, and laying it on his back carried it into his house and put it away.

A gentleman who came from India in the ship which brought the elephant Caroline, told me that the tediousness of the long passage was much re-

lieved by the interest they all took in this animal,
and the constant amusement she afforded them.
There was a great friendship between her and a
dog who stayed about her almost continually. At
the commencement of the voyage she was provided
with a covering, lest she should be chilled by the
sea air. But being still in the warm climate of the
torrid zone, she did not then feel the want of cloth-
ing, and immediately stripped off the garment and
threw it aside. Afterwards, when they proceeded
into a cooler latitude, and the covering was again
put on, she seemed very glad to have it, and wore
it without any further objection. At the termination of the voyage, the vessel encountered a violent storm, and was wrecked near the mouth of the
Delaware. The crew and passengers saved themselves in the long-boat. When they reached the
shore, they grieved exceedingly at having left the
poor elephant in the ship abandoned to her fate.
Some of the men volunteered to go back for her in
the boat, notwithstanding the terrors of the storm
and the imminent risk of their own lives. When
they reached the ship ; they found the elephant in
great tribulation ; but they could not prevail on
her to come away with them till she had provided
for the safety of her friend the dog, by taking him
in her trunk and handing him down to the boat.
This done, she gladly allowed the men to make her
fast by a rope to the stern of the boat, and thus

she swam after them to **the** shore. The elephant
Caroline was afterwards exhibited in Philadelphia.

Once, when she was thrusting her trunk about
among the spectators in search of something good
to eat, a young man mischievously gave her some
tobacco, which the elephant (not knowing what it
was) conveyed immediately to her mouth, but instantly put it out again with signs of the greatest
disgust, in which she showed her sense. A few
days afterwards, the same young man was there
again. The elephant directly remembered him,
and singling him out from the crowd, put forth her
trunk, and seizing the offender's hand, squeezed
and pinched it so hard as to make him cry out
with pain.

The tusks of the elephant supply the whole world
with ivory. **It is valued for its whiteness,** hard-
ness, and the fine polish of which it is susceptible.
There is a small insect, invisible to the naked eye,
which sometimes gets into articles made of ivory
and eats holes in them in a very ingenious manner.
Miss Leslie says : " I have a fan entirely of ivory,
which is almost as thin as the best white paper,
and is carved all over in a sort of delicate lace or
open work. It was made for me in Canton, when
a girl, and has in the centre the initials of my name
elegantly cut. In a few years the unseen worm
commenced his depredations, and my beautiful fan
is now eaten in small square holes of so regular a

form that they look as if made purposely with an instrument.   One of my sisters had a fine set of
• ivory chessmen that came from China, and after a while they were found perforated with small holes not larger than if pierced by a small needle. The chess-king was drilled completely through, from his crown down to his feet. It is probably a similar invisible insect that eats off the points of camel's hair pencils as they lie in the boxes at the stationer's, making them square at the ends, and therefore useless."

The spirited engraving represents a scene which took place in India, at a hunting-party, and is related by Captain Mundy, in his " Sketches in India," in the following narrative :

"A gentleman of our party had, perhaps, as perilous an adventure with a lion as any one ; he having enjoyed the singular distinction of laying for some moments in the very clutches of the royal quadruped. Though I have heard him recount the incident more than once, and have myself sketched the scene, yet I am not sure that I relate it correctly. The main feature, however of the anecdote, affording so striking an illustration of the sagacity of the elephant, may be strictly depended upon.

"A lion charged my friend's elephant, and he, having wounded him, was in the act of leaning forward in order to fire another shot, when the how-

THE ENCOUNTER.—THE RESCUE.

dah (which is the box upon the elephant's back)
suddenly gave way, and he was thrown over the
head of the elephant into the very jaws of the
furious beast. The lion, though severely hurt,
immediately seized him, and would shortly have
put a fatal termination to the conflict, had not the
elephant, urged by his mahout, or driver, stepped
forward, though greatly alarmed, and grasping in
her trunk the top of a young tree, bent it down
across the back and loins of the lion, and thus
forced the tortured animal to quit his hold! My
friend's life was thus saved, but his arm was broken
in two places, and he was severely clawed on the
breast and shoulders."

THE TRAVELED MONKEY.

"Oft has it been my lot" to meet,
Men of small wit, and large conceit,
Who, having visited, perchance,
The shores of Egypt, Greece or France,
Having seen the Pyramids, or sat
'Neath Sinai's shade, or Ararat,
Mounted St. Peter's or St. Paul's,
Or bathed beneath Niagara Falls—

Deem **their poor stock of** knowledge worth,
The hoarded wisdom of the earth,
And their mere dictum worthy quite,
To set all knotty questions right.

IN the central province, Chang Fou Tse, of the
Flowery Kingdom of the Sun, there was **an ex-**
tensive forest, remarkable for the magnitude **and**
beauty of its trees, the variety and richness **of its
flowers,** and the abundance of **its** delicious fruits.
But it was still more remarkable as the residence
of a tribe of monkeys, the most sagaciously human,
the most provokingly civilized, of any that **have
ever** been known to caricature the ways of man.
So exceedingly apt were they in learning the manners and customs of their more intelligent neighbors, that it was commonly remarked, that a more
vain, self-conceited, selfish race of thieves, pick-
pockets, and highway robbers was never known.
From the universal prevalence of these elements
of moral depravity among them, it was **currently**
believed by the philosophers of that age, that **this
was the identical family** from which **Lord** Mombo-
do, and others of the same school, traced their gen
ealogy **direct.**

If this could be satisfactorily proved, either **by**
authentic documents, or unquestionable tradition,
it would be a fact of great interest and importance
to the scientific world, as it would afford a natural
and easy explanation of certain psychological phe-

nomena, always exhibited in that family of philosophers. I refer to their singular talent for disputing everything that is certain, and believing everything that is doubtful. Nature has a less decided and unchangeable abhorrence of a vacuum, than these men of the simplicity and directness of truth. They will grope through weary volumes of misty speculation, and impalpable conjecture in quest of the "vestiges of creation," which, like foot-prints in the solid rock, are graven on every page of the volume of nature, and illustrated, in characters of light, in the book of revelation.

The tribe that occupied this beautiful forest of Chang Fou Tse, was known by the name of the Hing-po-qua tribe. They were large and well formed, with features more regular than others of their race. They were cleanly and social in their habits as well as exceedingly loquacious and communicative. Such a set of chatterers, babblers, and boasters, the sun, in his circuit, never looked upon. The imitative faculty, so strongly developed in the whole race, was pre-eminent in them, and was distinguished by a degree of refinement and taste elsewhere unknown. They were rarely caught doing anything ungenteel, or ungraceful, according to the paradoxical terms. On this account, they were often taken to the great cities of the east, and sometimes sent to distant lands, to be exhibited for the admiration of the curious.

It happened in the year — that one of the Hing-po-quas became the prisoner of Henry Cabot, a European merchant of great celebrity, who was acquainted with nearly all the countries of the globe and visited many different nations every year. The animal was tall and finely formed, with a coat so soft, smooth, and glossy, that his master gave him the name of Joseph Silk, to which, in acknowledgment of his remarkable gifts as a traveler, was added the surname of Munchausen. He accompanied his master wherever he went, and was received with marks of distinguished consideration, even in the most fashionable circles of society, being allowed the privilege of amusing gentlemen and ladies, by mimicking their movements, and caricaturing their looks.

On one occasion, being on a visit to the court of the queen of England, he was greatly delighted with the appearance of a very small page, in the service of one of the ladies of honor. He was quite a dwarf, and a good match, in point of size, for the monkey. Silky Joe, as he was more generally called, annoyed this miniature page exceedingly, by following him at all times, and acting over, with the most ludicrous precision, all his attitudes and motions. Happening, one day, to find the door of tne page's room open, as he was passing, the monkey stole in, and helped himself to a complete suit of court dress, of the richest materials, and of the

gay and showy colors so much admired among the
fashionables of that period. In attempting to array himself, he made some awkward mistakes.
Having satisfied himself, however, that all was
right, he hastened to take his place as the shadow
of the page. The court was in an uproar of laughter. The page was highly incensed, and demanded
satisfaction for the insult, vowing he would never
put on the dress again, after it had been so dishon-
ored. He was soon pacified with presents, and Mr.
Cabot purchased the dress at a high price. Induced, by this incident, to promote Joseph Silk to
the rank of a page in his own retinue, he also purchased for him a variety of other costly dresses,
after the most approved costumes of the day. He
had his head powdered according to the prevailing
custom, and his face bleached, by a liberal use of
depilatory appliances and pearl white.

Joseph was thus a frequent visitor at court, and
in many of the palaces of the nobility. He was in
universal favor, especially with the ladies, for whose
attentions he always showed a marked preference.
His manners were graceful and courteous in the
extreme. He could enter a room with the air of
an accomplished dancing-master. He flourished a
cane, an eye-glass, a pocket handkerchief, or a snuff
box, with the grace of a courtier. In fine, he was
a model of an accomplished beau, to whom brains
are superfluous. As to his tail that was no more

in the way than a Chinaman's queue. It was immediately adopted as a fashion, both by gentlemen
and ladies, the former attaching huge pig-tails to
their heads behind, and the latter making long trails
to their dresses.

After several years of absence, Mr. Cabot returned to the Flowery Kingdom of the Sun, and visited
the province of Chang Fou Tse. Joseph Silk Munchausen accompanied him, having become learned
in the manners and customs of men, and in the airs
and arts of travelers. Arrived within the precincts
of his native province, he was seized with a strong
yearning for the home of his youth, and an invin-
cible desire to astonish the natives, by showing off
his finery and his acquirements. Arraying himself
with great care in his choicest habiliments, as he
would have done for a presentation at court, he
seized an opportunity when Mr. Cabot was too much
occupied to notice his movement, and stole away,
with rapid strides, to the forest. Flourishing his
cane with vigor, he strode down the long avenues,
and through the favorite haunts of his childhood,
without encountering one familiar face. Supposing
him to be a human monster, "an outside barbarian,"
the Hing-po-quas, his cousins and neighbors, hid
themselves from him, as from an enemy. At length
one of them, peeping after him, from a hollow tree,
discovered his tail, half concealed by the ample
flaps of his coat. Stealing noiselessly out, he seized

the obtruding member, and gave it a violent pull,
as if to test its genuineness, and then, with a provoking chuckle, flew away into the tree. It was
answered by a hundred chattering voices in all its
branches. Gentleman Joe, though sorely offended
by the indignity offered to his person, instinctively
replied to the salutation, wheeling suddenly about,
and squinting scornfully through his eye-glass, to
see if he could detect the offender. The monkeys,
shrewd at all sorts of tricks themselves, readily suspected some trick on this occasion. They therefore, kept a respectful distance, chattering to each
other, and making all manner of grimaces at the
intruder, as, with the most affected airs imaginable,
he strutted about, sometimes brandishing his cane
in defiance, and sometimes threatening to chastise
them for their insolence. He soon became cool,
however, and, revealing his true character, invited
them to a parley.

One by one the monkeys gathered around the
stranger, till he had a large audience, to whom he
made himself known as a friend and relative, and
very condescendingly related the marvels he had
seen in distant lands. Doubt it not, kind reader—
Rousseau has settled that point long ago—that animals talk. The parrot, you know, is quite a lin-
guist, and talks Spanish, French, Dutch, German,
and whatever you may please to teach him. Other
animals have languages of their own, which have

never been reduced to writing. And, in this respect, I do not see why the brute is not entitled to
the pre-eminence, since many of them learn to understand, and some of them to speak, the various
human tongues ; while no man has yet been able
to learn any of their dialects. This question I
leave to the philosophers, and beg they will consider whether it is does not intimate, if not prove
clearly, that we, in our arrogance and self-conceit,
have mistaken the direction of the scale of being,
when we have placed man at the head of it.

Joseph Silk Munchausen was a plausible, insinu-
ating monkey, easy of address, and ready of speech
on all occasions. He was now especially desirous
to make a grand impression. He had a remarkably
happy faculty of showing up all he had seen, and
giving it the coloring of his own fancy. Some of
his hearers were as credulous as the disciples of
Swedenborg, Mesmer, or Mormon ; though there
were many unbelieving wags among them, who
made the most quizzical grimaces, as he dilated
upon some of the impossible marvels of other lands.
Monkeys are associationists. They **have** all things
common. And when Joseph told them that the
people of England allowed a few of their number
to call the whole land their own, and to claim even
the forests, so that their neighbors and brothers
could not so much as walk under their shade, they
laughed outright at their folly. " It is even worse

than that," said he, " there are a few who have
every year as much food as would serve ten thousand, which they keep all for themselves, while the
rest die by hundreds for want of a root, or a nut."

"Oh ! what a whopper," said one, " don't I know
that the hungry would take it by force, if it were
not given them ?"

"If they should do that, they would be hung,"
replied the traveler. "In truth it is only a grand
contrivance they have for getting rid of that mean
sort of people, that have not wit enough to make
themselves rich by wholesale robbery. They are
not allowed to have anything of their own. They
must either starve or steal, and if they steal they
are hung."

"Now we have caught you," said a grave old fellow, who had not spoken hitherto. "You say that
they hang those who steal. How could these rich
ones get so much land, and claim all the forests,
without stealing it ? It was not always thus. So
they should all be hung, and the land be common
again."

"You know nothing about human philosophy,"
replied Joseph Silk. "He who steals a morsel, to
save himself from starving, is a villain, and must be
put to death. But he who steals a whole kingdom
is a hero, and men worship and serve him, as a
kind of god."

A sort of suppressed titter, expressive of extreme

incredulity, was all the reply which the audience deigned to give to what they conceived to be a mere fiction of the speaker's fancy. He went on, however, in the same strain, slandering the poor humans with such malicious inventions as these:— " Let me tell you, moreover, that if any man kills another, the law is that he shall be put to death. But if he kills a hundred or a thousand, he is honored and rewarded as one of the greatest benefactors of the race."

" Caught again in your own trap," replied the philosopher—" for he who had killed one in a passion, would only have to kill a hundred more, and that would save him, and make a great man of him at once."

" What is the matter with your face, cousin ?" asked one, " that it is turned so pale and smooth? You have lost entirely that fine ebon complexion and hairy comeliness that we Hing-po-quas prize so much."

" No, no," replied the traveler, " white is the favorite color among men. So much so, that, in some places, it is deemed a crime to have a black skin."

" Impossible !" interrupted the sage before mentioned. " However, it is as well as could be expected from ourang-outangs without tails. I see who it is. The same judgment of heaven which deprived them of that fundamental ornament, deprived them also of their powers of reasoning. We must pity them, for they do not know any better."

" Tell that story to the Chop-picqs," cried one of the doubters, " Hing-poquas are not so easily imposed upon."

" 'Tis true," reiterated Munchausen, " true, every word of it, I have seen it with these eyes. The whites have nothing to do with the blacks, but to trample on them. They must not sit together, nor walk together, nor even pray together. I should never have seen any good society, much less should I have been a general favorite there, if I had kept my face black. Men dress their bodies in black, and think nothing so fine as black eyes and black hair ; but to have a black face is a sin."

With an air of perfect satisfaction, Joseph noticed the indications of surprise and disgust among the auditors ; for, like most other travelers, he deemed it greater honor to himself, to excite the wonder and tax the credulity of his hearers, than to secure their confidence as a truth-telling and honest observer. He, accordingly, went on with his story, always taking care to make his human brethren appear as ridiculous as possible. Among other incredible things, he had the audacity to declare, that a large majority of mankind were not at all scrupulous of right—that they practised lying, and thieving, and all kinds of injustice—that the

weak were everywhere oppressed by the strong,
the simple overreached by the shrewd, and the ignorant imposed upon by the shallow pretenders to
knowledge. He even went so far as to state, as a
veracious historical fact, that the larger portion of
mankind, of all ranks, were in the daily habit of
drinking various kinds of slow poisons, by which
they were often made sick, and by which vast mul-
titudes were annually killed. "It was not uncommon," he said, "for companies of them to meet together, with a view to see who could drink most of
such poisons. It always made them foolish, and
sometimes drove them mad, but still it is everywhere regarded as a sovereign remedy for all the
diseases it produces."

"That comes of their losing their tails," said the
philosopher.

"There is another custom," continued this veracious reporter, "which is almost too vulgar for a
refined Hing-poqua to believe ; but, I assure you,
on the word of a traveler, that it is, in a sense,
omniprevalent. They have a certain kind of very
dry, nauseous dust, which they are fond of, and
which they always eat with their noses, though it
almost invariably throws them into sudden and
painful convulsions."

A general shout followed this burst of original
wit, and the old woods rang again with the mirth
it occasioned. Nothing abashed, the modest speaker

waited a moment till the uproar had subsided, and then gravely re-assured his hearers that it was even so as he had said. "Why," said-he, "every *gentleman* carries a box of this singular powder in his pocket, offering it often to his friends, by way of salutation, as he meets them. And you will sometimes see a dozen of them together cramming their noses with it, and then, when the convulsions come on, shouting at each other, as if they were mad, and shedding tears, as if the operation were highly painful."

No sooner was this said, than one of those unbelieving wags before-mentioned, thrust his hand into his cousin's pocket, to test the truth of his statement. To his surprise, he there discovered a small black shining box, with curious figures on the top and sides. He opened it, and found it nearly filled with a coarse, brown powder which had a very disagreeable smell. Determined to try its quality, he took a handful of it, and crammed it into his nose, in doing which he was unfortunately not careful to keep it out of his eyes. Immediately his head seemed in a strange commotion. His eyes contracted, his nerves tingled, and he seemed about to swoon; when, suddenly and involuntarily, he uttered a convulsive shriek, that alarmed the whole audience, and sent them scampering away. The tears rolled down his cheeks, and he was left in that foolish predicament, that he did not know

whether he was laughing or crying, whether he
was hurt or pleased. The monkeys soon recovered
from their fright, and returned to inquire what was
the matter. They found the sufferer recovering
from his agony, the tears streaming down his face,
and mingling with the dirty brown powder, which
made him appear disgustingly filthy. He **was**
obliged to go to the brook **and** wash himself, and it
was long before his eyes ceased to smart, and **his**
nose to tingle under this unnatural stimulus.

This experiment was not without its advantages
to the traveler. It gained him credit among his
hearers, by confirming a part of his tale, which led
them to suppose **that the whole** might be true.—
He therefore went on boldly **to** say, that this **sin-**
gular powder was made of the dried leaves of a
plant, the various uses of which constituted some
of the chief pleasures of man. Sometimes they roll
up the leaves into a cylindrical shape, and then,
setting fire to one end of the tube, draw the smoke
into their mouths, and then puff it out again, till **it**
is all **consumed.** Sometimes **without** setting fire to
it, they put a large roll of it into their mouths, and
chew it. This mingling with the fluids **of the**
mouth, makes a very dirty mixture, which they are
so eager to get rid of, that they do not scruple to
spue it upon every object that is near them. It
often runs down upon their faces, and clothes ; and
I am compelled to say, that, in spite of all their

pride and self-esteem, men are more filthy and vulgar in many of their habits, than any of the animals I have seen."

"What could you expect of animals without tails?" asked the old sage, with the satisfied air of a victor. "I tell you, my friends, it is in this beautiful extension of the spinal column that the intellect resides. Therefore it is placed at the base of the column that it may sustain the whole. Therefore it is made ornamental, that it may attract the eye, and command the admiration of all. Therefore it is made flexible and pliant, that it may reach to every part of the body. And therefore, when that is gone, the poor unfortunate animal is reduced in the scale of being. He becomes a mere animal, and is like a ship without a rudder."

"True, true," interrupted Joseph, sneeringly, "but what do you know about ships?'

The philosopher deigned no reply, and there was a momentary pause.

"How it is," asked one of the company, "that your legs are so oddly shaped? They are round and smooth as the young willow twig."

"Oh! that is the fashion where I have been. In Europe, they have a class of men called tailors, who are constantly employed contriving means to cover up and hide the deformities of the human figure, though they sometimes make them more deformed than they were by nature. They were first em-

ployed in devising various kinds of long skirts, as
substitutes for tails, and this gave them the name
which they are known by to this day. A tailor is
the most important man in human society. He
holds all the rest in absolute subjection. And,
while a king rules over one country only, a tailor
rules over many—for all men must necessarily wear
the same kind of clothes, though they live in different parts of the world, and speak different languages, and are at deadly enmity among themselves.
These tailors change the fashion as often as they
please, and all men, everywhere, are obliged to
adopt it. If they neglect, or are not able to do so,
they are not considered worthy of good society.
When the fashion is close and tight, fat men are
obliged to squeeze themselves almost to death, in
order to get into their clothes. When they are
large, lean men are obliged to stuff them out with
cotton, or feathers, or any other convenient thing,
so that one can never know what a man's figure is
by his outward appearance. The tailor manages
that according to his own fancy or convenience."

"Poor, unfortunate race," exclaimed the sage,
" let us commiserate their infirmities, and be thank-
ful that, while our tails remain to us, we have no
need of tailors."

One by one, every article of the traveler's dress
was examined and commented upon, with numberless questions in respect to their various uses.

Seeing a sagacious old fellow puzzling himself with various experiments to detect the quality and use of an eye-glass, which was suspended by a chain from his neck, Joseph condescended to enlighten him, by assuring, that in good, that is to say, genteel society, it was not considered polite for one person to look at another with the naked eye, and glasses were invented, as a proper medium of vision for those who would show deference and respect to each other. An eye-glass was, therefore, a kind of passport to good society. And it was a singular phenomenon, that, as soon as any one rose from a low condition to good standing in society, he lost the power of seeing clearly without the aid of a glass.

Many other marvelous and ridiculous stories did the Hing-po-qua Munchausen relate for the diversion of his old friends and associates—of the manners and customs of different people, of their towns and cities, their palaces and ships, and numberless other matters, till they were weary of hearing what they could not believe. At length, with a general yawn, they bade him good night, and betook themselves to their several places of rest. Gentleman Joe, being greatly fatigued with his travels, and with his long effort to entertain his companions, was glad of the opportunity to retire to the hollow of an old tree, which was reserved for his special use.

As his custom was, he divested himself of all his clothes, carefully hanging them on the twigs and branches that grew about the door of his chamber. Having acquired the genteel habits of civilized society, his old friends were all awake and busy, while he was yet but half refreshed. At length, at a late hour in the morning, he opened his eyes, yawned, stretched himself, and turned over for another nap. In doing this, he caught a glimpse of his elegant embroidered coat running swiftly along the principal branches of a neighboring tree. Starting suddenly up, he found to his chagrin and utter dismay, that not one article of his finery remained. His wakeful cousins had borrowed it all, and there they were, one with his cap, another with his shirt, a third with one stocking on his leg, a fourth with another on his arm, capering and frisking about, with infinite glee and merriment. In vain did he attempt to recover them by giving chase first to one and then to another. He had lost a portion of his nimbleness and power of climbing, and his old friends only laughed at him for his present unavailing rage, as they had done before for his vain pretensions. Deprived of his trappings, and both ashamed and afraid to return, in a state of nakedness, to his master, this accomplished traveler was obliged to return to his original mode of living, and to the humble obscurity of a mere monkey.

THE CAMEL.

A RABIA is a large country of Asia ; there are few rivers in it ; there are few towns or trees, but there are a great deal of sand, and wide deserts. Only a few of the people live in houses, the greater number live in tents ; they have very fine horses ; they love their horses very much, and are very kind to them. The horses live with them in the tents, and never kick or hurt the children. Some of the Arabs are merchants ; some are shepherds, and some are robbers.

The merchants cannot carry goods which they buy, and sell, as we do in ships and boats ; because there are not rivers to sail upon in Arabia.

The Arabians have an animal which is very useful to them. This is the camel. He travels for
them, gives them milk, and his hair makes their
clothes ; he is of as much use to the Arabian as the
horse, the cow, and the sheep are to us : he is as
useful to him, as the reindeer is to the poor Lap-
lander. The camels carry loads of three or four
hundred pounds ; they kneel down to take up the
load, and rise when it is put on ; they will not allow more to be put upon their backs than they can
carry ; if more is put on, they cry loudly till it is
taken off. When they are loaded, the camel trots
about twenty-five miles in a day ; but when the
camel carries only a man on his back, he can tra-
vel one hundred and fifty miles in one day.

The camel drinks a great quantity of water at
once ; he has a safe place in his stomach, where he
can keep the water a long time, and when he is
thirsty, he wets his mouth by forcing up some of
the water. One sort of camel is called the drome-
dary. Some kinds of the camel have one bunch on
the back, others have two bunches. Camels live
forty or fifty years.

THE HAMSTER OR MARMOT.

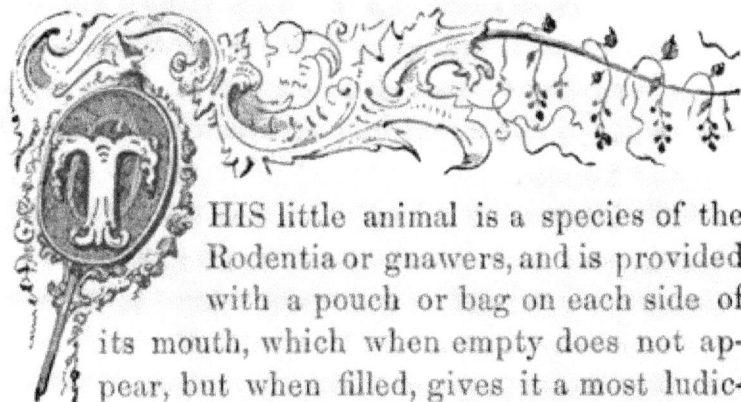

HIS little animal is a species of the Rodentia or gnawers, and is provided with a pouch or bag on each side of its mouth, which when empty does not appear, but when filled, gives it a most ludicrous and droll appearance, causing it to look very much as little boys look when suffering with the *mumps*. In these pouches the hamsters store their pilferings in the grain field ; and when they have packed away as much wheat or oats or rye as they can carry, they scamper off to their burrows or underground houses, and when they have unpacked from these natural receptacles one load of "stealings," they return to the fields after more, for they are among the veriest little commorants in the world ; and in this way, manage to plunder from the farmer a surprising number of bushels of valuable grain, which with their sharp teeth, they cut off ear by ear, carrying it unthrashed to their own neatly kept granary for their winter store.

This species of pouched rat is about nine inches long from its nose to its tail, the tail being about

three inches in length and with but little hair upon
it, resembling in this respect, the common rat.
The color of the hamster is a dark yellow, variega-
ted with black, yellow and white irregular spots.
It is sometimes found almost black in some cases,
relieved by lighter spots and with a white or yellowish muzzle.

The hamster is a very rare and beautiful little
animal, and is an inhabitant of Saxony, that part of
Germany now under the government of Prussia,
and has thus become identified with the birth-place
of Martin Luther, the Reformer, as it is seldom
found elsewhere.　Species of it, however, have
been captured in other parts of Germany, and
sometimes in Siberia and the southern parts of
Russia.

The hamster is very shy, but when attacked is
fierce and savage.　A favorite resort of his, when
hard pushed, is to jump on the breast or shoulders
of the hunter who corners him, and striking his
long sharp teeth fast in the flesh of his enemy, thus
supports himself while he tears and scratches him
most vigorously with his piercing claws.

The houses of the hamster differ in size according to the difference of their age.　The young ones
do not burrow over a foot under ground, whilst the
old ones often dig to the depth of five feet, and in
diameter, the habitation for each family is frequently ten or twelve feet.　The principal chamber,

which is the bed chamber of the old couple and their young family, is warmly lined with dried grass or moss. The other chambers of the habitation are used for storing provisions, and one is usually set apart for the use of the head of the family, he being either fond of the seclusion of uninterrupted retirement, or else willing to leave his wife sole mistress of the domestic arrangements without any meddlesome interference, which ladies usually so loudly deprecate.

Each habitation has two holes, one for ventilation and one for a passage of ingress and egress. One of these descends in an oblique direction, and the other perpendicularly. The young mature very quickly, and, like rabbits, they would increase in numbers with most amazing rapidity, were it not for the efforts of the hunter, who lays all manner of snares and traps for their destruction on account of the value of their fur.

The fur of the hamster, though coarse, is highly esteemed for cloak-linings, to which use it is for the most part appropriated. It is also used for other trimmings, and sometimes for ladies' muffs. It is eagerly sought by the trapper, and commands a good price.

As winter approaches, the hamster, who has taken such good care to store his cellars with provisions, retires into his subterranean abode to return no more to the upper air till spring. He carefully

closes the main entrance after him, and thus secure,
feeds and fattens upon his palatable grains, until
the great cold of winter comes upon him.   He then
rolls himself up into a ball, and sinks into a sleep
as profound as that of Rip Van Winkle himself,
though of not so long duration, the sleep of the
hamster lasting only while the cold weather lasts.
In this state of torpidity curious physiologists have
experimented upon the poor animal: They have
found the body cold and the limbs inflexible as
though with death ; the only signs of life, on opening the animal, being in the heart, which has been
found to pulsate very slowly ; so slow as to be
scarcely perceptible.

THE ALPINE HARE, or as it is sometimes called,
Alpine Marmot, is another species of the Marmot
family, though somewhat larger in size, being sixteen or eighteen inches long.   Linnæus and others
place it among the family of rats or gnawers, although some naturalists deny its close relationship
to the marmot.

The Alpine hare is considered one of the most
interesting animals of its whole class on account of
its habits, the beauty of its fur, &c.   In summer
the color of its fur on the upper part of its body is
of a grayish yellow or brownish ash, while in winter it is of a snowy white, like the Ermine, all over,
with the exception only of the tips of its ears, which
are at all seasons of a jetty black.

The Alpine hare inhabits the mountainous regions of Europe, more particularly those of Switzerland and Saxony, from which it takes its name.
It delights particularly in regions just below perpetual frost and snow, and in winter, instead of bur-
rowing in the ground, even when the storms are
most severe, and the cold most intense, it lives in
burrows made in the deep snow. And here, unlike
the marmot, it never falls into a torpid state, but
keeps up the high temperature of its body, even in
the severest cold. These burrows it generally
makes close beside the root of some small tree, or
upland bush, which, from being warmer than the
snow, melts it sufficiently around the bark to form
a sort of chimney or breathing hole for the sagacious dweller underneath. These snow palaces are
not, by any means, uncomfortable abodes ; snow being a non-conductor of heat, forms, in fact, a warmer shelter from the cold than a hole in the earth.
Then the white fur of the animal, which is warmer
than any other colored fur, prevents the heat of its
body from escaping, so that altogether, these imi-
tators of the Esquimaux have a very comfortable
life, and the little columns of smoke, which of a
clear cold day are seen arising from their breathing
chimneys, where there is a large colony of them,
appear not unlike a miniature settlement of the
snow huts of those Indians of the Polar regions—
the Esquimaux—with the smoke of their fires as-

cending from their odd shaped windows. In fine weather the Alpine hare loves to enjoy the sunshine, sitting outside its habitation upon its hind feet in an erect position, and looking abroad upon the scenery like any other amateur artist. But they always take the precaution to place a sentinel on guard, so that they need have no fear of a surprise. While they are engaged in eating, the sentinel keeps double watch, and at the slightest approach of danger gives a shrill whistle, and the whole army of eaters disappear into their burrows. These burrows are formed in this shape Y, like the letter Y. The Alpine hare has not the same facility for running that the common hare has, its legs being much shorter, but as the eagle is almost the only enemy it dreads, owing to its elevated place of abode, it has not the same need for fleetness of foot, that characterizes the common hare. The eagle it manages to elude by darting under cover, at the first cause for alarm.

The Histonwish, which is found in the wilds of the western continent, is another species of the marmot. There is also another, called the Quebec marmot, found in the northern parts of North America. This species lives mostly in trees, makes its burrows in dry spots, and passes a very solitary life.

MY SQUIRREL.

QUIRRELS are amusing little fellows. I wonder if any of my little Merry cousins ever tamed one. I have an idea floating around in that part of my cranium where the brains are supposed to find an abiding-place, that not many of them ever undertook the task; not that it was laborious, oh no; we of the Merry family never think of that, but because it was almost, if not quite, absolutely impossible.

But somehow or another I was rather fortunate in the taming of my squirrels. I am not bound to know whether there was anything very attractive about me, which induced "Bunnie" to place so much confidence in me or not; suffice it to say, he did, and no other shares the trust. Now for how I did it, for of course you all want to know that; but be patient, it will come in time; my pen is scratching away for dear life now, it won't go any faster.

One day last summer, and a warm day it was too, I took my usual walk to the brook, where, with a book, or perhaps my sewing, (oftener the former,) I've whiled away many a lonely hour, (quite ro-

mantic that, isn't it?) On this particular day,
however, I had a book, but it would not engage
my attention in the least ; do what I would, my
mind would wander, and in sheer desperation I
threw it away from me, and betook myself to the
delightful occupation of—tossing pebbles in the
brook. For not long, however ; for suddenly, just
above me, I heard a remarkably strange chirp.
"That's no bird," thought I ; and looking up, there
on a limb, just over my head, was perched the
prettiest little squirrel you ever did see, its bushy
tail curled up over his back, his fore paws holding
a nut to his teeth, and his little twinkling black
eyes dancing about with the most marvelous
celerity.

After watching him awhile, I ventured to call
"Bunnie." His eyes looked frightened, and glancing about, they at last came down to me. There
they stayed, till after satisfying himself as to my
identity, he scampered off, up one limb, across another, till out of sight! "How provoking !" I exclaimed, when a bright thought struck me. I wonder what kind of an effect "nuts " will have upon
his majesty? I pondered, and the result was a
tramp to the house and back, with a pocketful of
nuts. A few were laid on the ground beside me,
and I sat very quietly awaiting the result of my
experiment. I waited a long time before there
was any demonstration of Bunnie's presence, and

then I heard a slight rustle. I did not move ; then
another, and another, each one nearer, then glancing out of the corner of my eye, I saw Bunnie's
identical self close to my heap of nuts! I kept
perfectly still, and saw a little paw put out, which
clasped a nut, and away sped the little thief up the
tree, and away. Presently, back he came, and approaching my nuts, another was taken, and away
he scampered. And so he continued to come, each
time bolder than before, till there was but *one* nut
left! As he reached his paw for that, my hand
took the paw! A prisoner! and oh, so frightened!
"You'll be more comfortable in a minute, Bunnie ;
just look here!" and I held a nut before him. His
eyes glistened, and his little paw clutched it, like a
greedy little Shylock as he was. He cracked it, to
my infinite amusement and satisfaction, and picked
the meat out "beautifully," then looked up into my
face with a most "trustlike" expression in his little eyes for "more." He had another and another,
till after a while I ventured to release him, when
he perched himself on my shoulder, and there he
sat very contented, nibbling the nuts I gave him,
now and then looking into my face with an expression very like "gratitude."

Ever since, whenever I have been down, he always comes out to meet me, and when I sit down,
perches himself on my shoulder. Sometimes I talk
to him, holding him in my hand the while, then he

whisks his tail most understandingly, and looks *unutterables !*

Yes, my squirrel is, I do believe, the knowingest, cunningest, prettiest, and nicest little squirrel that ever lived. Now, shouldn't you think I'd love it? and love it, too, better than a dozen little tame playthings that always were, always are, and always will be, tame. Just you try it who can, and then see if you don't coincide with me.

A recent exploit of one of these sprightly and sharp-witted little creatures, belonging to a neighbor of mine, has so much interested and surprised me, as exhibiting passion, sagacity, and an obvious process of reasoning—so like the human race under similar circumstances, that I think it cannot fail to prove of some interest to others, and I therefore am induced to offer a brief relation of the fact for my numerous readers.

The squirrel in question, having been taken when very young, had become as tame and familiar as a kitten, and, up to the act by which he thought fit to sacrifice his home for the gratification of his resentments, he had shown himself quite amicable and harmless.

On the day of the incident about to be related, the owner having some company at the house whom he was treating with cracked walnuts, gave one to his pet. This being greedily devoured, the gentleman, by way of amusing himself and company, then selected a promising looking shell, carefully removed the meat, and putting the shell together again, placed it before his nut-loving favorite. The squirrel, never having been before deceived by a trick of the kind, confidently took up the shell in his paws, when perceiving it empty, he

let it fall with an air of evident disappointment. The experiment was then repeated. This was too much for the patience and equanimity of his squirrelship. On discovery that his second nut, thus insultingly offered him, was, like the former, destitute of the expected treat, he turned an angry glance

upon the author of the trick, and springing up,
seized him by the thumb, which he bit to the bone,
and then, though no word or blow was offered or
given, running out of the house, immediately retreated to the woods, from which he has never returned.

THE GREY SQUIRREL.

THE pretty Grey Squirrel lives up in the tree,
A gay little creature as ever can be;
But, though gay, he is prudent, and works **like the ant,**
To provide in the summer, for cold winter's wants.
He seeks out a hole in an old tree's core,
Where he makes a warm nest, and lays up his store
And when winter comes, and the trees are all bare,
And the white snow is falling, and keen is the air,
He heeds not the cold, as he sits by himself
In his warm little nest, with his nuts on the shelf.
O wise little squirrel! no wonder that he
In the green summer woods is as gay as can be.

THE LION.

THE lion is of a uniform gray or tawny color; the male, from his fourth year, has his head, neck, and shoulders covered with a mane; this gives him a majestic appearance, and distinguishes him from the rest of the feline tribe. His length, from the muzzle to the tail, is about five feet two inches; his tail, which is furnished at the extremity with a tuft of hair, is about two feet two inches in length. The female brings forth, at times, but one whelp, and never more than six. The whelps are born with their eyes open, and are as large as our domestic cats; at first their bodies are covered with brown stripes, running diagonally; their ears do not erect themselves until they are two months old; the mother defends them with terrible fury.

At the present day the lion is found only in **Africa**
and some parts of Asia : in old times he inhabited
Syria and Greece, between the rivers Nestus and
Archelaus. Next to the Asiatic tiger, and American jaguar, he is the most ferocious beast of prey.
He lives to a great age. In the year 1760, a lion
died in England, which had been confined in the
Tower for more than seventy years, and another
died there also at the age of three and sixty.

The following description will give us a correct
idea of the nature of the lion.

"The lion," says Lichtenstein, "like all the feline tribe, springs upon his prey, and never attacks a man or a beast that does not run from him,
without having first crouched to the ground at
a distance of ten or twelve paces, and measured
his leap. Hunters take advantage of this fact, and
it has become a rule with them never to fire until
he crouches, when at that short distance they can
take aim with such certainty, that the ball strikes
him exactly in the forehead. When a man is so
unfortunate as to encounter a lion unarmed, his only
hope of safety is in his courage and presence of
mind. If he attempts to run, he is infallibly lost ;
if he stands still quietly, the lion will not attack
him. He must not allow himself to be disturbed,
even if the animal approaches quite near him, and
crouches, as if about to take his leap ; he will not
venture this leap if the man has courage enough to

stand motionless as a statue, and look calmly in
his eyes. There is something in the lofty form
of man which inspires the lion with respect and a
distrust of his own strength, and the calm attitude
of the body increases this impression every moment.
It would be banished by an imprudent movement,
which should betray fear, or challenge the beast to
the attack. The result shows that his fear has been
no less than that of the man ; for after a while he
rises slowly, retires some steps, continually looking
back ; crouches again ; retires still, at shorter and
shorter intervals, and at last, when he thinks himself beyond the dangerous proximity, takes to
flight with all his speed. Unanimously as this fact
is asserted by the inhabitants of all parts of Africa,
yet the experiment can scarcely have been made
very often.

Formerly, when lions were in greater abundance
and the colonists had not learned how to hunt them,
they arranged a great hunt in common against a
lion ; tried to lure him into the open country, and
formed a large circle about him. If he tried to
break through on one side, they fired at him from
the opposite, and when he turned angrily upon his
new assailants, he was easily overcome by the numerous bullets which they discharged from all sides.
But he is now usually hunted by two in company ;
and skillful marksmen, who are sure of their aim,
and can depend upon their weapons, venture to go

alone in pursuit of a lion, and even to seek him out
in his lair. Such an undertaking is very dangerous, however, and many accidents have occurred.
Here are two examples.

The field commandant Ijaard Vanderwald, and
his brother Johannes, were pursuing, not far from
their dwellings, on the eastern slope of the Schnee-
berg, the track of a lion, which had done great mischief among their herds, and discovered him at last
in a ravine which was overgrown with thick bush-
es. They took their position on either side of the
entrance, and sent in their dogs to hunt the animal
out. They succeeded in this ; the lion rushed toward the side where the last named brother stood,
crouched, and received the contents of his musket.
Unfortunately the shot had not hit him directly,
but had just grazed his ear and the side of his
breast. After an interval of hesitation, which lasted for a few seconds, the beast recovered himself,
and darted, furious with pain, with such rage upon
the hunter, that he had scarcely time to leap upon
his horse and endeavor to fly. But in a few bounds
the lion overtook him, leaped upon the back of the
horse, which, crushed by the weight, could not stir
from the spot, and struck his claws into the thigh
of the unhappy man, seizing him at the same time
with his teeth by his clothes. While he clings
with all his strength to the horse, to prevent himself from being dragged to the ground, he hears

his brother galloping up behind him, and calls to him to shoot, hit whom he might. The brave Ijaard leaps from his horse, calmly takes aim, and shoots the lion through the head, and strangely fortunate, the ball passes through the saddle, without injuring either horse or rider.

Another man was not so lucky—a hunter called Rendsburg, who, with a cousin of the same name, went to hunt a lion. The adventure took exactly the same turn as the former one, but the lion leaped sidewise upon the horseman, and seized him with his teeth by the left arm. His cowardly comrade, instead of assisting the unhappy man, ran to call upon some men for help, who were posted not far off, at another outlet of the thicket.

In the mean while, Rendsburg had resorted to his last means of defence, and while the enraged animal lacerates and crushes his left arm, he draws with his right a knife from his pocket, and pierces the breast of the furious beast in several places. Those who had hastened to his assistance, found him dragged from his horse, swimming in his blood, his left arm entirely torn from his body, his side dreadfully lacerated, and the dead lion lying upon him, with the knife in his heart. After a few moments, the bold hunter, exhausted by loss of blood, breathed his last.

A writer, whose testimony may be relied on, tells us that, in many parts of the mountains, not far from Elephant river, lions are found in such abundance, that once, when on a journey, he saw two and twenty of them in one spot. The most of them were young, eight of them only being full grown. He had just unharnessed his horses upon an open place, and retreating with his companions to a distance, without venturing a shot, he gave up his beasts a prey to the wild beasts, who killed six of them, and dragged them away.

Near Rietrieviersport we came to the house of a man named Vanwyk. While we let our cattle feed a little, and sought the shadow under the portico of the house, Vanwyk related to us the following story : It is somewhat more than two years since I ventured a dangerous shot on the spot where we now stand. Here in the house, close to the door, sat my wife. The children were playing near her, and I was without at the side of the house, busied with my wagon, when suddenly, in broad day, a large lion appeared and laid himself quietly down near the threshold in the shadow of the portico. My wife, benumbed with terror, and aware of the danger of flight, remained in her place ; the children took refuge in her lap. Their cries attracted my attention. I hurried towards the door, and you can imagine my astonishment, when I found the passage barred in this manner. Although the animal had not seen me, yet, unarmed as I was, it seemed impossible to save them. I turned, however, almost involuntarily, toward the rear of the house, towards the chamber in which stood my loaded musket.

Fortunately I had, by chance, placed it in the corner nearest the window, and could reach it with my hand, for as you see, the opening is too small to allow me to climb into the room, and still more fortunately the door of the chamber was open, so that I could see the whole of the frightful scene. The lion now made a movement; it was perhaps about to take a leap. I hesitated no longer, whispered a word of encouragement to my wife, and fired, with a low "God help me." The ball passed close to my boy's curls, and struck the lion above his flashing eyes upon the forehead, so that he fell dead instantly.

It not unfrequently happens that the lion when he sleeps is awakened by the hounds, which are the constant companions of the caravans. Thus Barchell relates the following adventure: One bright day, at noon, as our dogs were diverting themselves by beating the reed-covered bank of a river, they suddenly broke out into a peculiar and loud barking; we sought for the cause of the clamor, and were soon convinced that they had seen a lion. We urged them on, and soon beheld a large lion, with a black bushy mane, and a lioness. We saw the latter but for an instant, she disappeared so quickly amid the reeds.

The lion, on the contrary, stood still, and gazed
steadfastly upon us. Our situation was not destitute of danger, for the lion was but a few paces
distant from us, and seemed preparing to leap upon us. The most of us were on foot and without
suitable weapons. But we had no time for fear,
and necessity required an attack in order to escape
one. I kept well upon my guard indeed, held my
pistol in my hand, with my finger upon the trigger,
and the rest who were provided with firearms did
the same.

But soon the dogs began to throw themselves between us and the lion, surrounded him and kept up
a violent barking. The courage of the animals
was truly wonderful; they approached closer and
closer toward the sides of the mighty beast, and
now threatened him in front, barking violently, and
without betraying the slightest signs of fear. The
lion, conscious of his strength, remained quiet, and
fixed his eyes only upon us. The dogs grew bolder and bolder, and ventured even within reach of
his mighty paws. He now became annoyed at their
din; a slight movement of his paw, and two of his
bold antagonists lay dead upon the earth. This
was done without the least exertion, and so rapidly that we could scarcely understand the result.
We fired at him, and a ball struck him beneath the
fore ribs, so that the blood flowed. He remained
for a while in the same position, and then retired.

At the commencement of the previous century, there was, among other animals in the menagerie at Cassel, a lion, which was remarkably tame, at least towards the woman who fed and tended it. This was so much the case that the daring woman, in order to excite the wonder of the spectators, often ventured to put, not only her hand, but even her head, within the animal's enormous jaws. She had often done this without the slightest accident occurring, and still the old and true proverb was at last verified, "He who goes into unnecessary danger perishes in it."

One day, as the woman entrusted again her head to his jaws, the lion snapped them to, and broke her neck, so that she died on the spot. This was doubtless committed involuntarily by the lion, as, unfortunately for the woman in this critical moment, he was compelled, tickled, perhaps by a hair of his mane, to sneeze. The result at least seemed perfectly to justify this supposition, for scarcely had he remarked that he had caused the death of his attendant, than the good natured and grateful animal became exceedingly sorrowful, laid himself down near the body, would not allow it to be removed from him, refused all the food that was offered him, and a few days after this misfortune died from grief.

THE ELK.

THE Elk, or Moose Deer, inhabits the northern forests of Europe, Asia, and America. It is generally larger than the horse both in height and bulk. Its horns are shed annually, and are of such magnitude that some have been found that weighed upwards of sixty pounds. The neck of the Elk is so short and its legs so long that it cannot graze on level ground, but must browse the tops of large plants and the leaves and branches of trees. It can step without difficulty over a gate that is five feet feet high. When disturbed it never gallops, but escapes by a kind of quick trot. None of the deer tribe are so easily tamed as this animal, which is naturally gentle ; and when he is once domesticated he manifests great affection for his master. The Indians believe that there exists a gigantic Elk, which can walk without difficulty in eight feet of snow, is invulnerable to all weapons, and has an arm growing out of his shoulder, which it uses as we do ours. They consider him as the king of the Elks, and imagine that he is attended by numerous courtiers. With them the elk is also an animal of good omen, and to dream of him often is looked upon as an indication of long life.

The elk frequents cold but woody regions, in the forests of which it can readily browse on the lower

branches and suckers of trees, its peculiar structure
rendering grazing an inconvenient and even painful action. In winter, when the snow sets in, and
when the wolves in particular, urged by hunger,
assemble in troops to hunt those animals which
they dare not attack. singly, the elks assemble in
herds for mutual protection and warmth in forests
of pines and other evergreens. These herds con-
sist of several families, the members of which keep
very close together. In the severest frosts, they
press one against another, or trot in a large circle
till they have trodden down the snow.

Their favorite food, when the winter proves severe, is the buds and bark of the buttonwood, birch,
and maple trees, &c. They browse against an as-
cent in preference to level ground, which, owing
to their long legs and short neck, they cannot
easily reach. In summer, to escape the torments
of gnats and other insects, they take to the water,
and swim great distances with ease ; and they are
able thus to gratify their appetite for aquatic
plants.

The elk is easily domesticated. It will follow its
keeper to any distance from home, and return with
him at his call. Hearne informs us that an Indian
at the factory of Hudson's Bay had, in the year 1777,
two elks so tame that when he was passing in a
canoe from Prince of Wales Fort, they always followed him along the bank of the river, and at night,

or whenever he landed, they came and fondled on him in the same manner as the most domesticated animal would have done, and never attempted to stray from the tents. One day, however, crossing a deep bay in one of the lakes, in order to save a very circuitous route along its bank, he expected that the animals would follow him round as usual, but at night they did not arrive ; and as the howling of wolves was heard in that quarter where they were, it is supposed that the elks were destroyed by them, for they were never seen afterward.

HOW APES CATCH CRABS, AND CRABS CATCH APES.

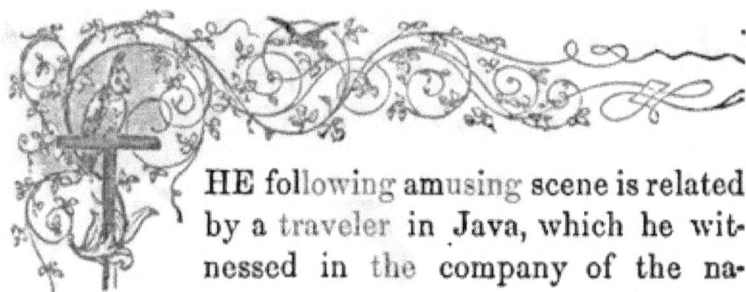

HE following amusing scene is related by a traveler in Java, which he witnessed in the company of the natives :—

"After walking close up to the old campaign, they were upon the point of turning back, when a young fellow emerged from the thicket, and said a few words to the mandoor. The latter turned with a laugh to Frank, and asked if he had ever seen the apes catch crabs. Frank replied in the negative, and the mandoor taking his hand, led him gently and cautiously through the deserted villages, to a spot which the young fellow had pointed out, and where the old man had formerly planted hedges, rendering it an easy task for them to approach unobserved.

"At length they reached the boundary of the former settlement—a dry, sandy soil, stript of beach, where all vegetation ceased, and only a single tall pandanus tree, whose roots were thickly interlaced with creeping plants, formed, as it were, the advanced post of the vegetable kingdom. Behind this they crawled along, and cautiously raising their heads, they saw several apes, at a distance of two or three hundred paces, who were pertly looking for something, as they walked up and down the beach, while others stood motionless.

"It was the long-tailed, brown variety, and Frank was beginning to regret that he had not his telescope with him to watch the motion of these strange beings more closely, when one of them, a tremendous large fellow, began to draw nearer to them. Carefully examining the ground, over which he went on all fours, he stood at intervals to scratch himself, or to snap at some insect that buzzed around him.

"He came so close, that Frank fancied that he must scent them, and give the alarm to the other monkeys, when suddenly passing over a little elevation covered with withered, reedy grass, he here discovered a party of crabs parading up and down on the hot sand. With a bound he was among them, but not quick enough to catch a single one ; for the crabs, though apparently so clumsy, darted like lightning into a quantity of small holes or cavities, which made the ground here resemble a sieve, and the ape could not thrust his paws after them, for the orifice was too narrow.

The mandoor nudged Frank gently, to draw his attention, and they saw the ape, after crawling once or twice up and down the small strip of land, and

peeping into the various holes, with its nose close
to the ground, suddenly seated himself very gravely by one of them, which he fancied most suitable..
He then brought his long tale to the front, thrust
the end of it into the cavity, until he met with an
obstacle, and suddenly made a face, which so amused
Frank, that he would have laughed loudly, had not
the mandoor raised his finger warningly. and directly the ape drew out his extraordinary line with a
jerk.   At the end of it, however, hung the desired
booty, a fat crab, by one of its claws, and swinging
it round on the ground with such violence as to
make it loose its hold, he took it in his left paw,
picked up a stone with the other, and after crack-
ing the shell, devoured the savory contents with
evident satisfaction.

 "Four or five he thus caught in succession, on
each occasion, when the crab nipped him, making
a face of heroic resignation and pain, but each time
he was successful, and he must have found, in the
dainty dish, revenge for the nip, and abundant
satisfaction for the pain he endured, or else he
would not have set to work again so soon.

 "Thus, then, the ape, quite engaged with the
sport, and without taking his eyes off the ground,
had approached to within about twenty paces of
the party concealed behind the pandanus trees.
Here, again, the ground was full of holes, and looking out the one he conjectured to be the best, he

threw in his line once, and probably felt that there was something alive within, for he awaited the result with signs of the most eager attention.

"The affair, however, lasted longer than he anticipated ; but, being already well filled by his past successful hauls, he pulled up his knees, laid his arms upon them, bowed his head, and, half closing his eyes, he assumed such a resigned, and yet exquisitely comical face, as only an ape is capable of putting on under these circumstances.

"But his quiet was destined to be disturbed in a manner as unsuspected as it was cruel. He must have discerned some very interesting object in the clouds, for he was staring up there fixedly, when he uttered a loud yell, left hold of his knees, felt with both hands for his tail, and made a bound in the air as if the ground under him was beginning to grow red hot. At the end of his tail, however, hung a gigantic crab, torn with desperate energy from his hiding-place, and Frank, who could restrain himself no longer, then burst into a loud laugh.

"The mandoor at first retained his gravity ; but when the ape, alarmed by the strange sound, looked up and saw men, and then bounded off at full speed, with the tormentor still dangling at the end of his tail, the old man could no longer refrain either, and they both laughed till the tears ran down their cheeks."

THE BEARS AND BEES.

A FABLE.

S two young bears, in wanton mood,
Forth issuing from a neighboring wood,
Came where the industrious bees had stored,
In artful cells, their luscious hoard,—
O'erjoyed, they seized, with eager haste,
Luxurious, on the rich repast.
Alarmed at this, the little crew,
About their ears vindictive flew.

The beasts, unable to sustain
The unequal combat, quit the plain ;
Half blind with rage, and mad with pain,
Their native shelter they regain :—
There sit, and now discreeter grown,
Too late, their rashness to bemoan,
And this, by dear experience gain,
That pleasure 's ever bought with pain.

So when the gilded baits of vice
Are placed before our longing eyes,
With greedy haste we snatch our fill,
And swallow down the latent ill.
But when experience opes our eyes,
Away the fancied pleasure flies ;—
It flies, but oh ! too late we find
It leaves a real sting behind.

THE GIRAFFE.

THE Giraffe, or Camelopard, has been long known to naturalists, though they have had but few opportunities of examining them in this country. They are found mostly in Africa, and are very docile and timid in their native state. What strikes you as most peculiar in looking at them is the enormous length of their neck and fore legs. The trunk of their body is short in proportion to their

neck. They are as tall as a small horse, and can
very easily see what is going on through the upper
windows of the building ; in fact, when they stretch
up their long, slim necks, it seems as if their heads
would be lost in the clouds. They feed upon the
leaves of trees mostly, as you see in the engraving,
though sometimes they take clover, barley, and
other grains. They chew the cud like the deer ;
resemble the camel in the length of their necks,
and the leopard in their spots. They are hunted
by the natives in Africa, for the sake of their large,
beautiful skin, and for the marrow of their bones,
which is considered a great delicacy. Some travelers have asserted that their flesh is a very good
article of food.

As there has been brought into our country two
very beautiful Camelopards, we believe the following account, taken from an English publication,
will be quite agreeable to our young readers.

The Camelopard, or Giraffe, although known to
the ancients, and captured for the purpose of adding to the barbarous exhibitions of Rome,* has
been rarely introduced to Europe in modern times ;
and scarcely ever, we believe, had Great Britain to
boast of a living specimen of the Camelopard within it, until the one arrived which was presented to

*Pompey, the triumvir with Crassus and Cæsar, had ten of these animals
at one time exhibited at the theatre, where wild creatures, as previously
stated, were let loose upon each other.

George IV., and which died shortly after its arrival.
—The largest preserved specimens, which have
been exhibited in England are, we are assured, the
two still to be seen at the British Museum, which
we should think are about fifteen or sixteen feet
high ; since they appear, in their preserved state,
much larger and taller than the finest of the living
specimens lately imported ; and of these no less
than seven have arrived in London within the last
two months, previously to the period of our writing. Of these there are four at the Regent's Park,
and three at the Surrey Zoological Gardens ; the
latter being considerably larger than the former,
and much more lively when we saw them. At
each of the menageries the Giraffes are attended
by three Numidian keepers ; but the highest of the
four which first arrived, does not, we are assured,
exceed eleven feet ; we have seen them only when
they were lying down, upon which occasions it was
said they were not to be disturbed.

Of the three which are at the Surrey Gardens,
two are males, and about from thirteen to fourteen
feet in height ; the female much smaller. Mr.
Cross informed us that the largest of the males was
fourteen feet high, two years and a half old, and
that the two others were about eighteen months.
The expense of the capture, and subsequent re-
moval of these seven animals to this country, must
have been very great ; and each party of them,

still attended by their Numidian friends, shows
they require more than ordinary care.

When we saw the three large ones at the Surrey
Gardens they were lively and playful as fawns, and
not in the least alarmed at the appearance of their
numerous visitors. They partook of their food, as
inclination prompted them, from a trough which
was placed at the height of an ordinary first floor
window.

The Giraffe, in a natural state, will grow to the
great height of seventeen feet, from the hoof to the
top of the head, whereon he has two small, erect,
and obtuse protuberances, like horns, which appear
to be covered with a tough skin. The shape of his
head is somewhat like that of the race-horse, yet
slender as the stag's ; his eye is dark and full ;
while his tongue is so peculiarly formed that he is
enabled to extend it a considerable length ; and by
encircling with it the tops of the light branches
and leaves of the trees upon which he feeds, he
thus obtains the chief part of his food.   His neck,
when he stands erect, is graceful and swan-like ;
his shoulders are high, and fore legs very long ;
the back slopes downwards from the bottom of the
neck to the insertion of the tail, which is thin, with
a tuft at the end. The fore legs are about two-
fifths of the creature's height, since they were just
six feet, in one which was fifteen feet high, as measured by Mr. John Campbell.

The body of the Giraffe is remarkably short, according to its extreme height, and is not the length of the neck from the top of the shoulder to the tail ; the legs are slender and have a tuft of hair on the knees ; the hoofs are hard and cloven, like other animals that chew the cud ; and the color of the skin is a light ash, or dun, marked all over with dusky red, or chocolate-colored spots. In their native solitudes they are hunted by the Arabs for their flesh—which is good eating—and their skins. They fly from the least noise, and ascend a precipice with the swiftness and security of the goat, the hoofs of both being similarly formed ; but though ever ready to retreat, yet, if closely pressed, this timid creature then uses its hoofs in its defence with the rapidity of lightning, and often not without effect.

M. Thibaut, who procured the Giraffes for the proprietors of the menagerie in Regent's Park, in a letter dated the 2d of January, 1836, and addressed to their secretary, says, "I availed myself of the emulation which prevailed among the Arabs ; and, as the season was far advanced and favorable, I proceeded immediately to Kordofan.* It was on the 15th of August, 1834, that I saw the first two Giraffes. A rapid chase on horses accustomed to the fatigues of the desert, put us in possession,

*Kordofan is a country of Africa lying to the westward of Nubia, or Sennaar.

at the end of three hours, of the largest of the two ; the mother of one of those now in my charge. Unable to take her alive, the Arabs killed her with blows of the sabre, and cutting her to pieces, carried the meat to the head-quarters, which we had established in a wooded situation ; an arrangement necessary to our own comfort, and to secure pasturage for the camels of both sexes which we had brought with us in aid of the object of our chase. We deferred until the morrow the pursuit of the young Giraffe, which my companions assured me they would have no difficulty in again discovering."

On the following day the party started at daybreak, " and at nine o'clock in the morning," says M. Thibaut, " I had the happiness to find myself in possession of the Giraffe." He is silent as to the means adopted in its capture, but says, " a premium was given to the hunter whose horse first came up with the animal ;" the chase having been " pursued through brambles and thorny trees."

He thus; proceeds :—" Possessed of this Giraffe, it was necessary to rest for three or four days in order to render it sufficiently tame. During this period, an Arab holds it at the end of a long cord. By degrees it becomes accustomed to the presence of man, and takes a little nourishment. To furnish milk for it, I had brought with me female camels. It became gradually resigned to its condition, and was soon willing to follow, in short stages, the route of our caravan.

" The first Giraffe, captured at four days' journey to the southwest of Kordofan, will enable us to form some judgment as to its probable age at present, as I have observed its growth and its mode of life. When it first came into my hands, it was necessary to insert a finger into its mouth, in order to deceive it into a belief that the nipple of its dam was there ; then it sucked freely. According to the opinion of the Arabs, and to the length of time that I have had it, this first Giraffe cannot, at the utmost, be more than nine months old. Since I have had it, its size has fully doubled."

As we cannot imagine that any European can be better qualified than M. Thibaut, to speak of the habits of the Giraffe, we quote the following passages from his description of them. He tells us that its first run is so exceedingly rapid that the swiftest horse, if unaccustomed to the desert, would scarcely come up with it. If it reach a mountain, it passes the heights with rapidity ; its feet—as already observed—being like those of the goat, endowing it with the dexterity of that animal ; and with such incredible power it bounds over the ravines, that horses cannot, in such situations, compete with it."

The Giraffe is fond of wooded country, where, as we have said, leaves of trees are its principal food ;

its conformation allowing it to reach the tops.' **The
one spoken of above,** killed by the Arabs, measured
twenty-one (French) feet from the **ears to the** hoofs.
Green herbs are very agreeable **to** this animal, but
its structure does not admit of its feeding on them
in the same manner as our domestic quadrupeds,
such **as the** ox and the horse. It is obliged **to**
straddle widely ; its fore feet are gradually stretched apart from each **other, and its neck being then**
bent into a semicircular **form, it is thus** enabled to
collect the grass ; but on the slightest noise **the**
timid animal raises itself with rapidity, and has re-
course to immediate flight. It eats with delicacy,
taking its food leaf by **leaf**; and, unlike the camel,
rejecting thorns and coarse **herbage.**

M. Thibaut obtained **five** Giraffes at Kordofan ;
but, owing to the cold weather of December, four
of them died, leaving him with only the one which
he had at first procured. He, however, persevered,
remaining three months in the desert ; and at length
captured **three others, all smaller than** the **one**
which, it may be fairly said, he bred by hand.
With these four he has, after all his toils in an inhospitable desert, safely arrived in London. There
are three males and a female ; and, having shown
the trouble—the expense attending which may be
imagined—of obtaining these living rarities, it will
readily be supposed that less of both could not
have been spared in procuring those, which are so

much larger, in the possession of Mr. Cross, and which need not, therefore, be narrated.

In its present domestic state, "as the grass on which it is now fed," adds Mr. Thibaut, "must be cut for it, it takes the upper part only, which it chews until it perceives the stem to become too coarse. Great care is necessary to its preservation, especially cleanliness. It is extremely fond of society, and is very sensible ; I have observed one of them *shed tears* when it no longer saw its companions, or the persons who were in the habit of attending to it."

In conclusion, M. Thibaut adds, that the Giraffes in his possession were " capable of walking for six hours a day without the slightest fatigue ;" which, in growing animals, shows the great strength they must possess when in their full vigor. Those in the Surrey Gardens were fifty-six days in coming over to England. We have already spoken of the graceful appearance of the neck of the Giraffe, when he stands erect ; but we cannot particularize any other part of his form as corresponding with it ; so far from this, indeed, he appears altogether a more awkward animal than many with which we are better acquainted. In his wild state, and flying over the wastes of an African landscape with the swiftness of a hunted roe, he may appear, if not a symmetrical, yet a beautiful object in the distance. But we cannot admit that it is the elegance of his proportions, or the gracefulness of his movements, which render him so great an object of attraction among us.   The action of his walk, trot, and canter, appear very awkward ; the more especially, in consequence of moving both legs on the same side at the same time, and not transversely as do other quadrupeds ; and to the comparative shortness of his body—at every step the hinder foot not following the fore in a direct line, but passing it on the outside, and reaching considerably beyond it.

It is a rarity which deservedly renders the Giraffe an object of attraction among us, while his gentle nature at once confirms all favorable impressions which might have been previously created in his favor ; and, certainly, the great expense, enterprise and perseverance employed to obtain these productions of the desert, both in a national and scientific point of view, ought to, as we have no doubt it will, be liberally compensated.

THE WITCH RABBIT.

ABBITS were always great pets with me, either as a favorite around the house, in the garden, or served up on the table. We had gone out upon our morning walk, and as I was not in very good health, Mr. W. carried his gun along for the purpose of shooting a young squirrel or rabbit, which would be more dainty food for my palled appetite. We went over the brook into a bushy field, covered with a thick growth of hazel and brambles, with here and there a large old tree left standing.

We were pushing our way along a narrow path, when, hearing a sudden rustle among the dried leaves, we saw the white tail of a rabbit go bobbing up and down as he went bounding off. He did not go far, however, but stopped in sight, just in the thickest of all the hazel clumps.

We could see him through the slim stems, standing erect, with his great wide eyes staring at us, just as you see him in the cut.

"A fine young rabbit! and he will make you a nice meal!" said Mr. W., as he raised his gun to fire. Bang! went the gun; but Mr. Rabbit did not even wink.

Now, Mr. W. prides himself particularly upon being one of the best marksmen Kentucky has ever produced. "Why," he exclaimed, with an expression of mortification, "miss a rabbit not more than twenty paces off! how ridiculous! Why, the creature is not even scared! Wait a bit, my little man, and I'll see if I can not scare you some!" and he proceeded rapidly to re-load, when to his disappointment he found that in getting over the fence he had lost the stopper to his shot-bag, and all his small shot was gone, except about half a dozen. But he happened to have a few rifle-bullets in his pocket, so he put two of these into his gun along with the half dozen small shot.

Mr. Rabbit, in the mean time, stood as immovably staring at us as if he had been some goblin statuette, hewn from brown stone, with great ebony set eyes, and placed in some green nook to mock the passing sportsman; while the elfin creatures peeped from out the flowers, and clapped their hands in the mockery of tiny glee.

Bang! went the gun again! and the white spots showed themselves along the hazels in a direct line with the creature. But not a motion did Mr. Hare make!

"How strange!" said Mr. W. with a perplexed air. "You see the hazel stems are literally riddled

all in a line with the creature, and yet he does not
stir ! Can it be possible that he has died in that
strange attitude ? That is absurd ! However, we
will try him again !" and down went two more bullets, which were the last.

Bang ! went the gun again. No more white
spots appeared upon the hazel stems, but Mr. Rabbit remained still immovable. I could not help
laughing at Mr. W.'s humorously perplexed look
as he exclaimed—

"Well, the creature must be a Witch Rabbit,
surely ! I have no more shot, but I am not going
to give it up so ! Here's a hazel stem which will
about fit the bore of the gun, and they say that
hazel is deadly to witches, so I will shoot it at him.

Bang ! went the gun once more, and the rabbit remained, if anything, more immovable still. Mr. W. stared a moment, and laughing said—

"I am afraid this witch is too strong for even the spell of the hazel! You go round to the brook, and get me a handful of the small gravel over which it runs. I have frequently shot birds with it when my shot gave out."

I clambered the fence, and was soon back with the gravel, and to my great amusement, as well as astonishment, saw Mr. Rabbit still standing there, with his great wide unwinking eyes staring at us.

Mr. W. fired several times with the gravel, with the same result as before, until we both burst into shouts of laughter, which proved to have greater effect upon our Witch Rabbit than all the roaring of the gun, and he went bounding slowly off through the thicket ; and as we turned to go, we caught a glimpse of him a short distance off, staring after us with the same immovable stare.

We had a good laugh over the incident as we returned home. Mr. W. accounted for the quaint incident by saying he had probably lost the shot of the first charge out of his gun, by the loosening of the light wad without his being aware of it—for it will be remembered, there were no white marks after the first shot. The other shots with the two bullets, although in a straight line, were glanced by the thick hazel stems. The sticks were also

glanced, and the gravel we found, on closer exami-
nation, to be of so light a quality, that it could only
be propelled a few feet from the muzzle of the
gun.

We returned the next morning, and found our
witch in the same "form," and Mr. W. shot him
easily, running, at the first fire.

Thus ended the mystery of the Witch Rabbit,
which a superstitious person would have insisted
all his life in regarding as a supernatural event.
It might finally have become a legendary wonder.
Such is, undoubtedly, the origin of the marvelous
tales which fill the early literature of all lands con-
cerning weird animals.

THE RABBITS.

" I wish that you would come and see,
What Johnny Taylor offers me,
   Two rabbits small and white !
Do let me keep them in the yard :
I'll feed them well and be their guard,
   And nurse them day and night.

Do not say 'no,' my dear papa,
They shall not plague you nor mamma,
   For I will keep them clean ;
How very happy they should be,
If they, poor things, belonged to me ?
   Such beauties ne'er were seen !"

"Susan, could you in comfort dwell,'
Within a dark and narrow cell,
    Confin'd by bolts and locks?
Or can my darling girl suppose,
Those rabbits e'er could feel repose,
    Shut up in yonder box?

To keep them thus would be unkind,
For they by nature were design'd
    To ramble wild and free.
Then send them to their hills away:
There let them scamper, frisk and play,
    Enjoying liberty."

THE GOOD DOG AND BAD BOY.

O guard his store at night, my neighbor keeps a noble Newfoundland dog. Not long since I was passing his store at mid-day, when he came out with Towser at his heels and a pail in his hand. He told Towser to take the pail and carry it to the house, a few rods across the way. The dog did not whine over the command, nor curl his tail and refuse to go ; no, not he. He obeyed at once, took the pail in his mouth and away he went to the house. I watched him to see how well he fulfilled his master's orders. The door was closed, so he sat down on the piazza and waited a welcome. Five minutes passed, and no one opened the door ; yet the dog was patient and faithful. Five minutes more passed, and just as I was about to leave, he was seen from the window and admitted with his charge. Faithful dog, thought I, never to refuse obedience, or wait for the second bidding.

Then I thought of little Willie S———, who said to his mother in my presence, "No, I can't do it ; let Ned go—he is not doing anything."

"Willie," exclaimed his mother in a commanding

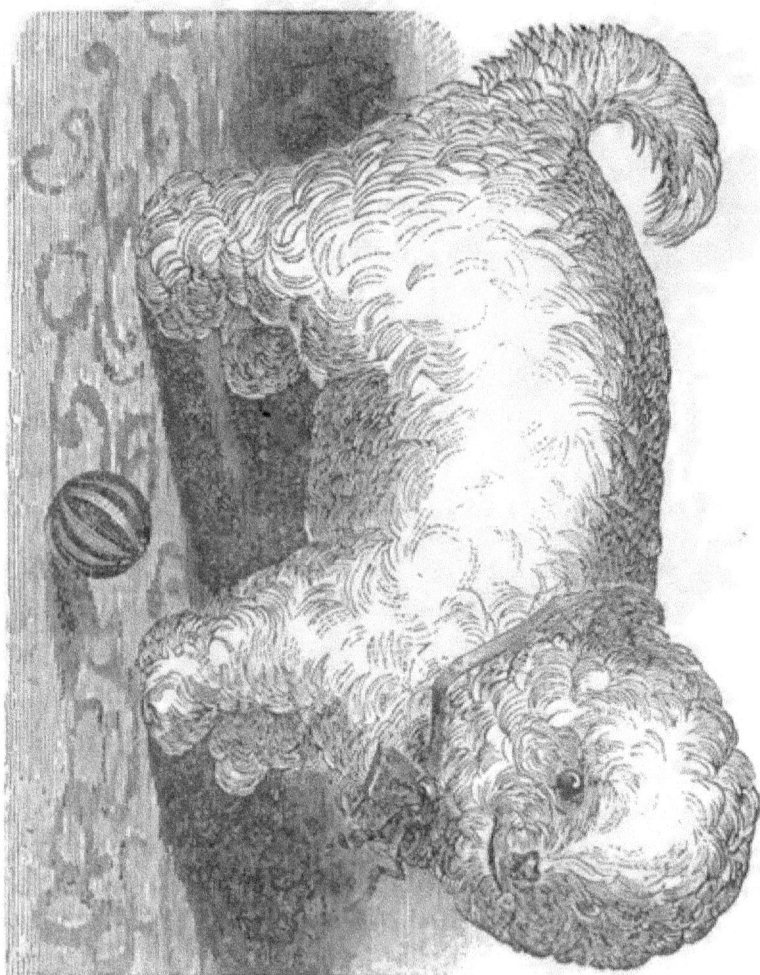

tone, "go and bring that wood immediately ; don't let me have to tell you again."

The little fellow was mending his cart, but he dropped his hammer, now that he saw there was no escape, and started. "I always have the wood to bring," he muttered as he left the room. He obeyed very reluctantly. He went pouting and murmuring after the wood, and when he returned he threw it into the box with a violence that threatened to break it to pieces. His mother looked ashamed and heart-sick. I pitied her from the depths of my soul. Think of it. Her son was less obedient than the dog ; for the dog went cheerfully, wagging his bushy tail, and lifting his head, as if to say, " *I obey.*"

Learn a good lesson from the example of the dog, and never let it be said of you, "Towser is more obedient than Willie."

———

"FATHER," said a cobbler's son as he was pegging away at an old shoe, "they say that trout bite good now."

"Well, well," replied the old gentleman, "you stick to your work, and they won't bite you."

A REMARKABLE CAT.

A CAT, which had been long remarked as one of the wildest of those which frequented a barn on the borders of a wood in Ayrshire,—so wild, indeed, as to be seldom seen,—was several times, during a sharp frost, observed, with no little surprise, to pass and repass into the adjacent farm-house, which it had not, for some years, been known either to enter or approach. It might have been inferred that it was compelled by hunger, had not this been the best season for catching birds; but, in one of its stealthy visits, it was seen snugly coiled up beside a baby in the cradle, to the no small horror of the mother, who imagined, in accordance

with popular prejudice, that it had come to suck
the baby's breath. All that could be done to per-
suade her of the impossibility of the cat doing this
was of no avail, and orders were immediately given
to every servant on the farm to kill the poor cat
wherever she could be found. Her caution and
agility, however, were long successful in saving
her ; and, though the persecution she thus experienced rendered her, if possible, much wilder than
before, yet she was not thereby deterred—not
even after being wounded by a pitchfork, and her
leg lamed by throwing a hatchet at her—from pay-
ing a daily visit to the baby in the cradle, because
it was the warmest place within her knowledge ;
and, next to food, she considered warmth as indis-
pensable to life. She persisted thus in venturing
to the cradle, till she was at length intercepted and
killed.

THE HORNED OWL;

OR

NEVER TRUST TO APPEARANCES.

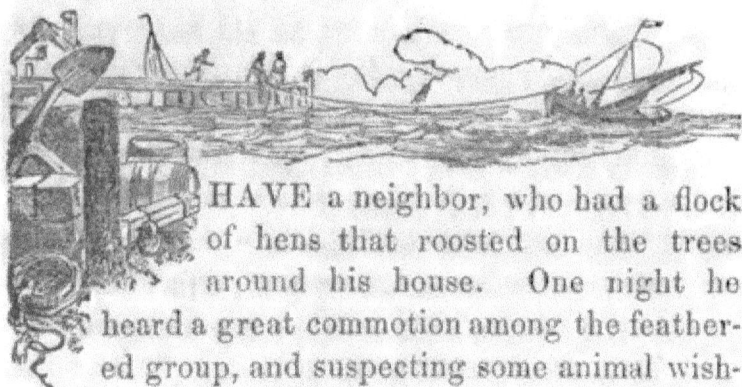

HAVE a neighbor, who had a flock of hens that roosted on the trees around his house. One night he heard a great commotion among the feathered group, and suspecting some animal wished a fowl for his breakfast, he took a gun and went out. Sure enough the depredator was there, and he supposed by its looks that it was a barn-yard owl with the chicken in its claws, just making its exit, without stopping to say, "Good-bye." The man fired, and the thief dropped to the ground, quitting his hold upon the chicken, and helpless himself with a broken wing.

On inspection he proved to be a great horned owl, a species rather rare in this region. As a curiosity, he was taken into one of the village stores for exhibition.

A group of boys were collected around him, but rather afraid of his owlship, even in his disabled state. As they stood at a respectful distance from the bird, a gentleman, remarkable for his love of animals and his dislike of boys, came into the store. "Ah! indeed, a horned owl! a great curiosity! What are you afraid of, boys? No animal will ever hurt you if he is properly treated. There, now, my good fellow!" he said, pushing aside the boys, and laying his hand upon the mottled plumage of the bird, whose drooping wing and downcast look made him appear like a pining captive, "I am sorry for you." The bird no sooner felt the pressure of the hand than he started, threw out his large sharp claws, inflicting a wound upon the gentleman's hand, which made him regret his misplaced confidence for some weeks.

This same gentleman came into my garden once, as I was looking at the sun through a smoked glass, during an eclipse. It was at the moment of greatest obscuration, when there was a hush in all nature, as if the pall of death were about to be spread over the earth. My own heart was full of awe and wonder, and I was thinking of the desolation which would follow if God should withdraw the light and heat of the sun wholly from us, when a voice near said, hastily, "I must go up to the pasture quick, for my cows may be afraid."

I turned, and he was making his way out of the gate as fast as possible, while my husband who stood near with a glass in his hand, was smiling as he watched his hasty departure.

REYNARD, THE FILIBUSTER.

A TERRIBLE fellow was Reynard for stealing,
  A trade he pursued without conscience or feeling;
He cared no more for a crying hen
Than the pestilence cares for suffering men.

He would creep right up to a well-set roost,
And help himself to what he liked most ;
And would tear young chickens right out of the arms
Of screaming fathers and fluttering ma'ams.

Or, leaving the chickens to shirk as they could,
He would tear the fond mother away from her brood,
Making no more count of the " family figure,"
Than a sheriff would do in seizing a beggar.

What chickens to orphanage suddenly brought,
What ducklings or geese in extremities caught;
What sensitive fowls rendered childless, or *widdered*,
This hard filibuster in no wise considered.

He prowled in the barn-yard, he skulked in the hedge,
Wherever through crevice or crack he could wedge;
He was sly, he was shrewd, he was cunning and prudent,
There were some things he could do, and some things he couldn't.

He could run, he could hide, he could skulk, he could fly,
He had to his safety a vigilant eye;
But—there's always a but, soon or late, for the sinner—
He had ventured too far to be always a winner.

He had fattened on chickens, and ducklings, and geese,
Till the fattest and fairest were quite common-place
And daintly seeking a dinner more rare,
He had poached on the park and abstracted *a hare*.

· This was reckoned too much for the gentry to stand,
'Twas a crime—'twas a trespass—the laws of the land,
Which made nothing of common folks' chickens and hens,
By statute protected *gentlemen's* pens.

The fever was up. Poor Reynard was doomed,
A vagabond, fugitive, outcast, presumed
To have no condition but that of a thief,
From whom the said gentry demanded relief.

A hunt was got up by the true law-abiders,
There were all sorts of horses, and all sorts of riders,
Determined to get on the track of the rogue,
And make him, if found in his hole, disembogue.

Now a hole Reynard had, 'neath the roots of a tree,
He thought none would ever discover but he,
A hole so peculiarly guarded without,
You would say, should you see it, 'twas safe, without doubt.

And then, as if taking the hint from the creature,
Whom the dog Noble pestered with bark, (*vide* Beecher,)
The tree that protected his cabin was hollow,
And furnished a chamber where no dog could follow.

THE FOX.

Well, the hunters turned out, and day after day
They scoured the beat in the usual way:
But Reynard, accustomed by moonlight to scout,
Stayed snugly at home while the hunters were out.'

But finding, one night, a new batch of fat chickens,
And relishing greatly the delicate pickings,
He staid out so late the rare feast to consume,
That morning o'ertook him while hurrying home.

So a hunter, who started with earliest light,
Observed the sly rascal, and cut off his flight;
He sounded his bugle, and soon the whole pack
Of hounds, boys, and huntsmen were down on his track.

Suffice it to say—that Reynard, the cunning,
Had eaten so much it affected his running;
And ere he had finished two miles of the chase,
The hounds overtook him and cut short his race.

THE OPOSSUM.

HUNTING THE OPOSSUM.—The hunting of the opossum is a favorite sport with the country people in Virginia, who frequently go out with their dogs at night, after the autumnal frosts have begun and the persimmon fruit is in its most delicious state.

The opossum, as soon as he discovers the approach of his enemies, lies perfectly close to the branch, or places himself snugly in the angle where two limbs separate from each other. The dogs, however, soon announce the fact of his presence by their baying, and the hunter ascending the tree discovers the branch upon which the animal is seated, and begins to shake it with great violence to alarm and cause him to relax his hold. This is soon effected, and the opossum attempting to es-

cape to another limb, is pursued immediately, and
the shaking is renewed with greater violence, until at length the terrified quadruped allows himself
to drop to the ground, where hunters or dogs are
prepared to despatch him.

Should the hunter, as frequently happens, be un-
accompanied by dogs when the opossum falls to the
ground, it does not immediately make its escape,
but steals slowly and quietly to a little distance,
and then gathering itself into as small a compass as
possible, remains as still as if dead. Should there
be any quantity of grass or underwood near the
tree, this apparently simple artifice is frequently
sufficient to secure the animal's escape, as it is difficult
by moonlight, or in the shadow of the tree to distinguish it ; and if the hunter has not carefully observed the spot where it fell, his labor is often in
vain. This circumstance, however, is generally attended to, and the opossum derives but little from
his instinctive artifice.

After remaining in this apparently lifeless condition for a considerable time, or so long as any
noise indicative of danger can be heard, the opossum slowly unfolds himself, and creeping as closely
as possible upon the ground, would fain sneak off
unperceived. Upon a shout or outcry in any tone
from his persecutor, he immediately renews his
death-like attitude and stillness.

If then approached, moved or handled, he is still

seemingly dead, and might deceive any one not accustomed to his actions.   This feigning is repeated as frequently as opportunity is allowed him of attempting to escape, and is known so well to the country folks as to have long since passed into a proverb.   He is playing "*possum*" is applied with great readiness by them to any one who is thought to act deceitfully, or wishes to appear what he is not.

THE LYNX.

AMONG the several species of the Lynx, some are found in Asia and Africa, with black tips on their ears, which make it a very conspicuous animal. It lives on small quadrupeds, and birds, which it pursues even to the tops of trees. The Lynx has never been tamed—always when confined in a cage, it snarls at all who approach. The face resembles the cat's, which animal it seems to come near in the link of beings. It is larger, however, always being about two feet long, and more than a foot in height. The Canada Lynx has longer and

more curly fur, or almost hair; and is remarkable
for its gait.   Instead of walking, it always bounds
from all four feet at once, with the back arched.
It feeds principally on the American hare.   It is
about three feet long.   The natives eat its flesh,
which is white and firm, and much like the flesh of
the hare.   Its skin is an important article of com-
merce, and many thousands are yearly exported.

THE AGES OF ANIMALS.

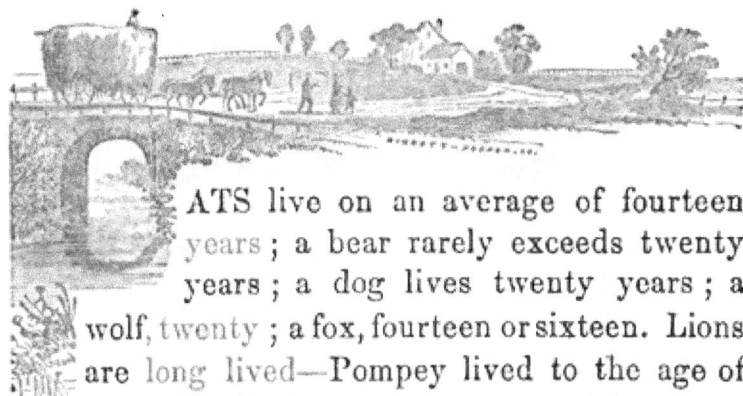

ATS live on an average of fourteen years ; a bear rarely exceeds twenty years ; a dog lives twenty years ; a wolf, twenty ; a fox, fourteen or sixteen. Lions are long lived—Pompey lived to the age of seventy. A squirrel and hare, seven or eight years ; rabbits seven. Elephants have been known to live to the great age of four hundred years. When Alexander the Great had conquered one Porus, king of India, he took a great elephant which had fought valiantly for the king, named him Ajax, and dedicated him to the sun, and let him go with this inscription : "Alexander, the son of Jupiter, had dedicated Ajax to the sun." This elephant was found three hundred and fifty-four years after. Pigs have been known to live to the age of thirty years ; the rhinoceros to twenty. A horse has been known to live to the age of sixty-two, but averages twenty-five to thirty. Camels sometimes live to the age of one hundred. Stags are long-lived. Sheep seldom exceed the age of ten. Cows

live about fifteen years ; Cuvier considers it probable that whales sometimes live to the age of one
thousand.   The dolphin and porpoise attain the
age of thirty.   An eagle died at Vienna at the age
of one hundred and four years.   Ravens have frequently reached the age of one hundred.   Swans
have been known to live three hundred and sixty
years.   Mr. Mallerton has the skeleton of a swan
that attained the age of two hundred and ninety
years.   Pelicans are long-lived.   A tortoise has
been known to live to the age of one hundred and
seven. ·

THE Arabs say, that every race of animals is go-
verned by its chief, to whom the others are bound
to pay obeisance.   The king of the crocodiles holds
his court at the bottom of the Nile, near Siout.
The king of the fleas, lives at Tiberias, in the Holy
Land, and deputations of illustrious fleas visit him
on a certain day, in his palace, situated in a beautiful garden, in the Lake of Genesareth.

THE IBEX.

THE Ibex inhabits some of the mountainous regions of Europe and Asia. It is sometimes
found among the Alps and Pyrenees. It is esteemed a great prize by the hunters, who give them
chase with great eagerness, and run into perils and
hardships innumerable to procure them. The hoofs
are very strong and sharp, securing them a firm
footing on the rocks they inhabit. They are expert leapers, and the time between the leaps is so
short, that the animal resembles an elastic sub-

stance, instead of a living creature renewing its
efforts at every leap. When pursued, they take to
the glaciers, along which they bound with great
rapidity, clearing chasms of a good many feet,
though, in general, they do not resort to such places
as they furnish no food.

Altogether, the Ibex is an interesting animal,
and it is not the less so that it is found only in the
wildest and most inaccessible places, and being
sought for at the extreme peril of the hunter. One
almost regrets their success, notwithstanding their
boldness, for the animals are very near extirpation.

The Ibex resembles the goat in form, but the
head is smaller in proportion. The eyes are large,
round and brilliant, and even fiery in expression.
The horns are flat ; they incline backwards and
downwards.

DON'T KILL THE BUTTERFLIES.

HOW delightful the early morning walk! How bracing the bright air of a fine October morning especially! And it seems to me as if *children* more than others enjoy the early walk. Such was my conclusion, at least, on the particular morning whereon hangs my little story. How the sun did shine! how meekly did the harebell bow its delicate head, quite unable to look up, because of its pearly jewels! How beautiful the contrast between its pale azure and the fine yellow of the ladies' bed-straw! Such are a few of the common things so delightful to an observing eye, which met and lured us on in that walk. Nor must I forget the butter-flies—blue, white, pink, brown, and gold—flitting about in the enjoyment of their brief, bright life. No wonder that the children felt happy ; no wonder that it was thought " time enough for school yet!" And now, in the midst of all this beauty, pause awhile—hark!—hush! What did Edwin say? Again was the clear voice of the little boy distinctly heard—too distinctly for her who had undertaken the mental training of the hitherto misguided one.

" Kill it! kill it! There, I have it!"

" No, Edwin, you *must* not, *can not* kill so beautiful a creature," said the lady before alluded to.

" Oh, yes, I always kill them! See how soon I can bring it down!" and away he ran, hat in hand, intent only on the destruction of the happy fluttering insects which beautified his path.

No more was said just then, as by this time both the butterfly and its pursuer were far away. Presently, however, was heard the cry of victory.

" I have killed it! I have killed it! It led me a chase, but I brought it down at last! There it lies!"

"Yes," said the lady, " there it lies !—its painted,

velvety wings lie in the dust ; no more will it wing
its way through the delicious morning air, and delight you and me when we again walk out. Edwin,
do you feel happier now that you have taken away
the life given to that little creature by *its* God and
yours ?"

The little boy hung his head, but made no reply ;
and school-time having now fairly arrived, he was
left to his own thoughts, ănd the lady to hers ; the
latter not being without hope that the "word
spoken in season" might prove to be good.

Again was preparation made for the happy "going out." "I wonder, Edwin," said the lady, "if
the butterflies have trimmed their feathers this
morning ?"

"Feathers !" said the little boy, in evident astonishment ; "*birds* have feathers—*not* butterflies !"

"Well," said the lady, "I am only sorry I have
no microscope, which would show you that what
appears like fine dust scattered on their wings, is in
reality *feathers* of delicate texture, and each furnished with a quill, strange as it may seem to you.

Moreover, the wings are finished by a fringe, of more exquisite manufacture than ever adorned the royal robes of any other than a fairy queen. On the outside of its little eye-ball are placed no fewer than forty thousand perfect lenses, or, if you will, *little eyes,* which seem to say, so much beauty had need to be well guarded."

The boy opened his blue eyes a little wider than usual, but still said nothing. But why did the butterflies flit unmolested by Edwin in that walk? Why was the little straw hat unlifted, and the hitherto ready arm unraised? And who is it that stands earnestly beseeching a group of boys to *spare the butterflies?* It was Edwin. Need I speak of the lady's feelings at these happy results? or need I say how much more of real pleasure that little boy himself now feels since he has learned to admire and *spare the butterflies?*

THE HORSE.

ORSES are among the no-
blest of animals if not the
most useful to man. In a
wild state they are found
in large droves, numbering
sometimes a thousand or
more. Powerful as they
are, however, they never
attack other animals, but
content themselves with
acting on the defensive.   When they lie down to
rest, they generally leave some of their number as
sentinels, to give notice of the approach of danger.
When the alarm is given, by a loud neighing of the
sentinels, the whole troop start to their feet, and,
after taking a view of their enemy, either give
them instant battle, or gallop off with inconceivable
speed.

When they determine on battle, they close round
the enemy on all sides, and trample him to death.
If the attack is of a very serious character, they
form a circle, in the centre of which the young are
placed with their mothers. The rest arranging
themselves with their heels towards their foes, re-
pel the most vigorous attacks.

Many a careless boy, and unfortunate man, knows,

from bitter experience, what a powerful instrument
of defence the horse possesses in his heels.

The most beautiful horses in the world are the
Arabian, though there are different races of them,
as well as in other countries. The most remarka-

THE HORSE.

ble and valuable among them, are the Kochlan,
who, to an uncommon gentleness and docility, and
a singular attachment to their masters, unite a
courage and intrepidity worthy of the best trained
war-horse. They have an astonishing power of re-

membering the places where they have been, and the treatment they have received.

The intelligence of this race of horses is almost incredible. He knows when he is sold to a new master, or even when his old master is bargaining to sell him. When the proprietor and the purchaser meet for that purpose in the stables, the Kochlan appears instantly to guess what is going on. He becomes restless and dissatisfied; casts frequent angry glances from his beautiful eye at the merchant, paws the ground impatiently with his feet, and exhibits other unmistakable signs of discontent. Neither the buyer, nor any other stranger, dares to come near him. But, when the bargain is concluded, and the vender, taking the Kochlan by the halter, gives him up to the purchaser, and turns away, the horse becomes immediately tractable and submissive. From that moment he is mild and faithful to his new master, as he had been to his old one. This is no idle story. It is well attested by English residents in the East, as well as by Turkish, Arabian, and Armenian merchants.

We can hardly wonder at the extreme gentleness and docility of the Arabian horses, when we consider how they are treated. The Arabs live constantly in tents. These they always share with their horses. The mare and her foal occupy the same corner where the children sleep, and often

serve them for a pillow.   They may often be seen
prattling to their colts as our children do to their
pet dogs, patting them on their necks and faces,
stroking down their soft hair, climbing on their •
bodies, and hanging about their necks, with the
fondness and fearlessness of childhood.

The Arabian horses are always well fed, and
never whipped.   The use of the lash is not known
among them, and it is only in the utmost extremity
that the spur is used, and then as sparingly as possible.   They are seldom, if ever overburdened, or
overworked, but are treated with as much care and
tenderness as any member of the family.